AF291466

BECOMING ECOLOGICAL

BECOMING ECOLOGICAL

Navigating Language and Meaning for Our Planet's Future

Derek Gladwin and Kedrick James

AEVO UNIVERSITY OF TORONTO PRESS

Aevo UTP
An imprint of University of Toronto Press
Toronto Buffalo London
utppublishing.com
© Derek Gladwin and Kedrick James 2026

ISBN 9781487561864 (cloth)　　　ISBN 9781487561888 (EPUB)
ISBN 9781487561871 (PDF)

Library and Archives Canada Cataloguing in Publication

Publication cataloguing information is available from Library and Archives Canada.

Cover design: Tamara Hawkins
Cover image: iStock.com/Yuuji

The manufacturer's authorised representative in the EU for product safety is Mare Nostrum Group B.V., Mauritskade 21D, 1091 GC Amsterdam, The Netherlands. Email: gpsr@mare-nostrum.co.uk

We wish to acknowledge the land on which the University of Toronto Press operates. This land is the traditional territory of the Wendat, the Anishnaabeg, the Haudenosaunee, the Métis, and the Mississaugas of the Credit First Nation.

University of Toronto Press acknowledges the financial support of the Government of Canada, the Canada Council for the Arts, and the Ontario Arts Council, an agency of the Government of Ontario, for its publishing activities.

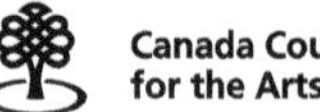

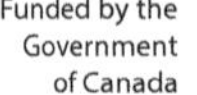

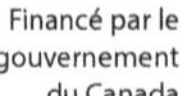

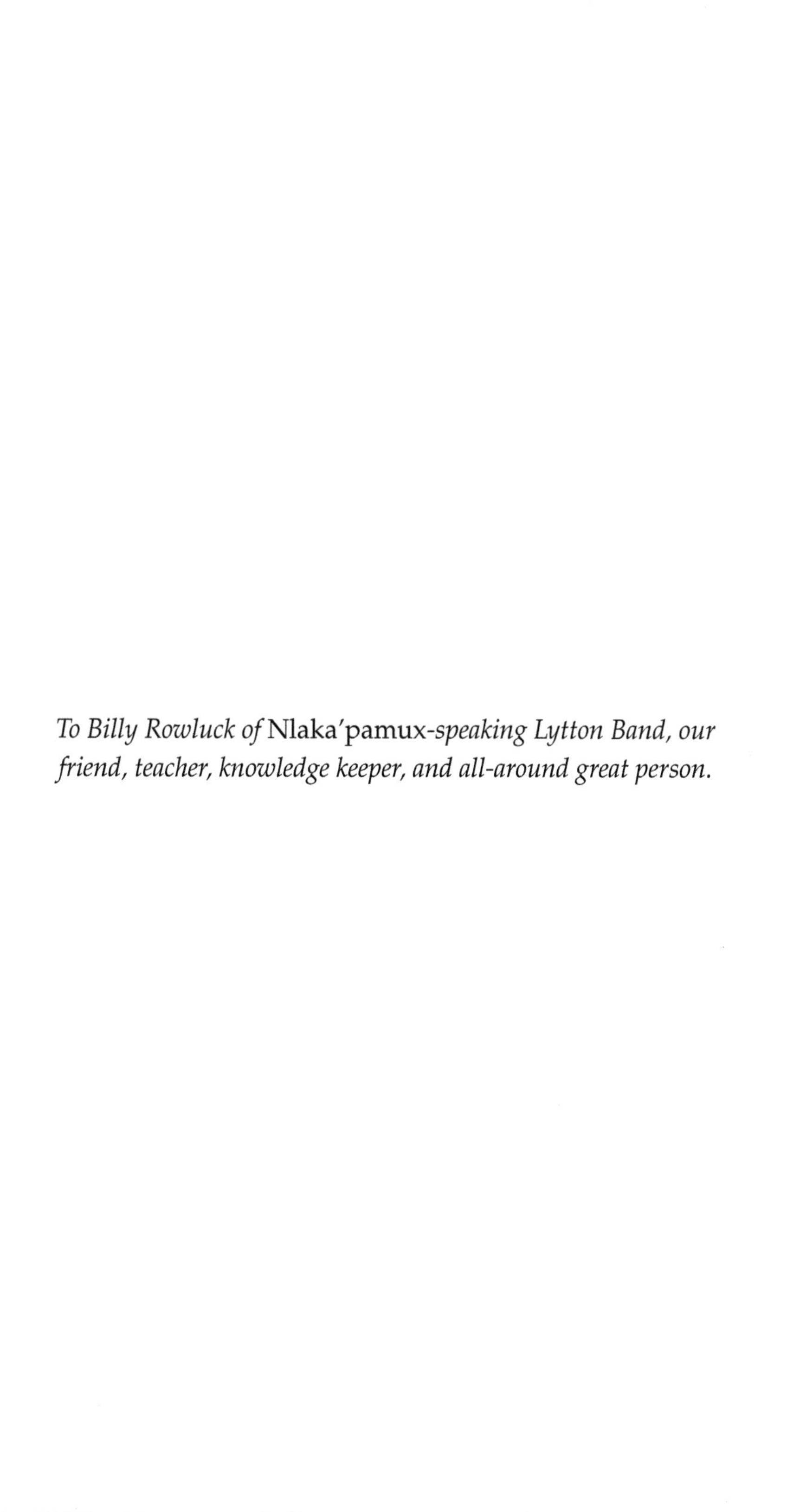

To Billy Rowluck of Nlaka'pamux-*speaking Lytton Band, our friend, teacher, knowledge keeper, and all-around great person.*

Contents

Prefacing

Adding to the conversation

This book was borne out of a desire to address, even in some small way, one of the most significant challenges not only in our time, but perhaps also throughout human history:

In a planet facing ecological collapse, how can we turn things around?

This is a big task, admittedly, but that's exactly why we're taking another approach, one that mirrors ecological patterns in both content and format. The goal is to spark a shift in how we understand and interact with the world. At the end of the day, we're all ecological anyway, so perhaps it's not such a huge ask after all. At the same time, we're exploring how society and people have direct influence in empowering these changes through the ways we communicate and create meaning.

So maybe there's a more tangible and relatable path forward: language lies at the heart of how we approach and comingle with the environmental challenges of the modern world. The power and influence of language shapes our perceptions, defines our priorities, produces our politics, and ultimately guides the actions we take (or fail to take) toward the kind of ecological future we need to both survive and thrive.

Another way to approach this question is to suggest that we're exploring how the ways people *talk* can lead to ways of *acting* with greater ecological awareness and responsibility. The philosopher and teacher Jiddu Krishnamurti often emphasized that the path through growing crises begins with internal *awareness* – a view of reality that frees us from judgment, rigid categories of right and wrong, and the impulse to criticize ourselves or others. In his words, "it is a deep psychological revolution that is necessary."[1] This inner shift, expressed through how we speak and relate through an ongoing process of *becoming*, can be the foundation for more conscious and transformative ecological action through awareness. Ecological discourse plays a huge role in this process.

Language serves as a facilitator between our inner awareness and the external, material environment in need of care. Through language – in all the ways we communicate, make sense of reality, and create meaning – we shape how we see the world and how we choose to engage with it. It holds the power to influence how we co-create a society rooted in ecological responsibility and collective well-being.

This viewpoint can, we believe, be a helpful guide for embracing a more generative path for our planet's future.

Some backstory

As the authors, we're inevitably situated in our own *language ecology* that may need some context. The way people approach conversations, along with the sources and examples we choose, shapes the language ecosystems we navigate and contribute to.

This book draws on a range of diverse perspectives, including Western, Indigenous, and Asian traditions; metaphysical philosophy; human–nonhuman relationships; ecological practices; multisensory experiences; dynamics of language; and multiple intelligences, among others. At the same time, the sheer number of possibilities means that many other valuable viewpoints had to be left on the side for other future conversations.

As professors of language and literacy education at the University of British Columbia, situated on the traditional, ancestral, and unceded Musqueam and Syilx territories, we engage with these concepts as both environmental educators and researchers. We're aware that all forms of knowledge and learning can quickly become complex, so we remain humbled by the realization that everything can't be covered. Accuracy and representation, much like language itself, continue to be elusively evolving in this journey. We can hopefully provide some helpful engagements with topics that we've been grappling with for years, as both educators and lifelong learners, and those we've found most useful in our classes and ongoing research.

But we also bring the concepts ahead to the forefront as living beings *sharing* this planet with other life. Educating for the changes we need involves rethinking what it even means to learn how to live ecologically. Who might we turn to for help?

Is this a priority for the earth sciences like ecology or biology? What about physics? Or education? Psychology? The arts? Or perhaps philosophy? There's also history, engineering, and so forth. Such an integrated topic – exploring how everyone is already ecological (but may not quite know it yet) – defies typical categories of expertise and definition. It further complicates categories of culture and identity as well. Just imagine how difficult it is for our publisher to find the proper "section" to place this book for booksellers.

As one might expect, there's somewhat of a backstory that might help to describe this more clearly. Or, well, to at least convey how we got here. Over the past several years, we tasked ourselves with developing a class on "environmental literacy" that would educate both teacher candidates and established teachers to not only *know* more about ecology, climate, sustainability, and so forth but also to *live* consciously with them, to understand and embody them beyond just knowing.

The aim was to explore questions like: how do we become ecological not just in mind but also through deeply felt, embodied practice? How can we pass this understanding on to future generations? These questions then emerged into: how do we recognize ourselves as ecological beings who are always in the process of learning and unlearning? How, then, might we communicate and practice this? Well, as we discovered, these questions quite often returned to discourse and how we use language as something that creates meaning in our lives.

This journey began when we served as sustainability fellows along with other educators across the university in various disciplines. We had monthly conversations about implementing several courses and curricula focused on sustainability, climate, and ecological education. As we developed our own class on environmental literacy, we wanted to revise outdated curricula and co-design something that we ourselves, our students, or even our colleagues would be excited to learn more about. Or, to be more accurate, that would have lasting benefits for everyone as co-habitants of the planet.

Designing and co-teaching this course over a few years led us to embrace a systemic pedagogy – an interconnected ecology of ideas – rather than the traditional linear approach to curriculum, with a fixed start and end and a direct path between them. We kept wondering how could the conditions of education shift how we perceive the world around us? The answer initially came clearer to us in the form of language, catalyzed through the emphasis on verbs as framing devices instead of nouns, illuminating the possibility that these ideas aren't static but shifting as environments around us change and extend into other deeper aspects of knowing and being.

Although not explicitly focused on teaching or pedagogy, this book serves as a broader call to education in the more

expansive sense of citizen-education and lifelong learning –
an *education* of values and ideas, and meaning that extends
beyond what we might be otherwise limited to a *schooling*
of tools and practices. It's about the experience of being a
part of living ecological systems and what we call *becoming*,
as a process of ongoing learning and creative growth that's
always in flux.

This approach encourages shifts in perspectives and
worldviews, ranging from minor adjustments to significant transformations in how we perceive multiple parallel
worldviews or *ontologies* – addressing the nature of reality
and our understanding of what exists in the world around
us. These are fundamental to becoming: talking, thinking,
and acting ecologically.

It's clear, if one chooses to see it, that the Western values we've inherited, those rooted in what's often referred to
as "modernity," aren't working so well. Many even believe
that these values are in massive decline, due to the falter-
ing democracies and rising oligarchies all around the world.
The dominant Western worldview, which elevates personal
gain through competition and wealth accumulation, often
relies on exploiting and oppressing others and the environment for profit and power. It's built on the idea that things
follow simple cause-and-effect logic and that we can and
should control everything in order to prosper.

This pervasive mindset makes it harder, but not impossible, to create real change for society and the environment.
These circumstances have led us to reevaluate our approach
to how we acquire, understand, and shape knowledge, also
known as *epistemology*, and to ask how shifting our thinking

and language might influence the ways people act and find meaning in those actions.

What's also encouraging is that there are many other valuable worldviews and knowledges, many of which are rooted in teachings that address the fundamental interconnection among humans, as one species among many other living beings. One of our colleagues and friend Vanessa de Oliveira Andreotti calls this *depth education* – recognizing how certain knowledges have been privileged over others and responding by creating space for diverse worldviews.[2] As settler inhabitants of these lands, where the histories of claimed ownership are disputed, we recognize that true relational responsibility in depth education involves reassessing established beliefs that have contributed to the current state of socioecological crisis.

Settler-colonial practices that claim ownership based upon *individual human* rights, as opposed to rights for *all living beings*, could certainly use a do-over. And such ownership isn't only about land; it also encompasses the power and knowledge of language. To take this seriously, we all might enact paradigm shifts in what Potawatomi biologist and scholar Robin Wall Kimmerer promoted as a language revolution to shift our ethical responsibility with the ecological world.[3]

To address this ecological shift through language and meaning, this book explores ideas, word origins, teachings, storytelling practices, concepts, and metaphors that are rooted mainly in the English language. By playfully riffing on words or meaning to explore ecological realities, some of the book leans toward a poetic or philosophical practice rather than academic or scientific approach. As language and literacy educators, we want to take some creative license with language and the meaning it creates, stepping outside conventional academic frameworks while also working in English, the dominant Western language often used in mainstream environmentalism.

About the approach ahead

Regardless of this backstory about how our educational experiences brought us here, this isn't only a book about educating and learning, nor is it limited to our influences and curiosities. It's not even about any one topic, really, but about the many things that intersect with ecological possibilities all mediated through language and meaning. Put somewhat poetically, it's an *experiential tour* of sensations, perspectives, and adaptations – an exploration of possibilities and *matterings*, or the ways that our relationships with matter, or the material world, matters. As the title suggests, it explores some possible aspects of becoming ecological.

Many books and articles available to us about environmental and climate-related issues are often focused on problems and critiques, or *what isn't working*, and get into a trap of saying the same thing again and again without any proposed ways out of blame or cycles of frustration. Or, worse, they leave readers paralyzed with fear of the future through a cascade of alarming facts and statistics.

These doomsday scenarios, while containing a lot of potential truths, often halt the very social attention that's needed. We also get embroiled in this tendency too and sometimes fall prey to it ourselves. Ecological collapse on the scale we are witnessing is no laughing matter. One way out of this trap involves a shift in structure and approach, empowering people to think about and foster daily experiences and choices that make real change.

One of the debilitating aspects of this looming disaster approach is that most people get overwhelmed quickly and stop seeing the point of making incremental change in beliefs and habits to promote sustainable ways of speaking, thinking, and acting. This often leads to apathy about creating change in society, which circles us back to everyday language and how it shapes meaning in our lives, regardless of the outcomes that follow.

Will anything I or we do even matter at this point?

We're guided by language to make assumptions about the world and then possibly act. Each word or phrase we use has multiple connotations and thoughts associated with it. Those that reference qualities of goodness in the environment like *pure, natural, eco, green, pristine, sustainable, renewable,* and so on make us feel good about what we're doing. The opposite is true of words that describe qualities of badness for the environment, such as *toxic, polluting, impure, single use, destructive, collapsing,* and *wasteful.*

We're dealing with something quite complex – layers of goodness and badness, mixed with needs, desires, habits, and a nagging conscience trying to urge us to do better. And we need language that recognizes this complexity in our lives. We might consider doing something to become *less* polluting, *less* wasteful, *less* destructive, even if we know that we aren't reaching "zero waste" or becoming pollution-free. In this manner, we're reminded of consequences without succumbing to fatalistic inevitability and pessimism.

For many it's a terrifying thought that various systems we take for granted are likely to collapse in our lifetimes and that it's not one system at a time but rather a cascade, because each system relies on other systems. This thinking haunts people of all ages. It's humbling that with so much capacity for thought and feeling, humans have carelessly brought the planet to a tipping point for all other species. Humans can't seem to stop ourselves from tipping over. We seem to have lost the capacity to act across time, to live for a planetary future. We've lost the initial storyline of existence.

Perhaps there are still ways to collectively prevent systemic collapse, something that politics, science, and technology can't solve for us, something rooted in how we do and don't communicate with, and about, the environment?

Just to be clear, we arrive at this point with degrees of cautious optimism. We believe that, with several billion people alive on the planet, if every one of us could be more conscious about our connection to the ecologies that surround us, every day, toward sustaining life on this planet, it would very quickly start to add up. Why? Because it would influence other people on an exponential scale. But this vision would entail a certain degree of consciousness, a momentum that embraces what it is to be alive, to be aware of what living within our means *means*.

And, highlighting an ongoing theme throughout the book, if everyone one of us recognized that *we are always becoming ecological* (but perhaps don't fully realize it yet), then our systems of governance and industry would be forced to change apace. Becoming is a collective response already in process, so we might just need to look and listen more closely.

The looping characteristic of topics in this book – each chapter feeding off others, and the use of actionable and adaptable verbs as guiding chapter titles that don't fit easily into prescribed definitions or categories – invites readers to play by exploring and activating curiosities for ways we all might reconsider how language and communication are integral to how we make meaning in and of the ecological world where our lives are inextricably bound.

Chapters can be read out of sequence – although the first chapter, Discoursing, serves as a helpful introduction to some key concepts and foundational ideas about language and ecology. All the chapters provide a vast range of terminology, theories, examples, stories, and experiences that try to shift the ways we use language and communication to create generative meaning together. In these cases, we italicize words and provide some working definitions.

Also included are thought experiments ("Experimenting with Possibilities") and practical prompts ("Practicing"), plus reflection questions ("Reflecting") after each chapter and dialogue prompts ("Group Dialogue Questions") at the end of the book for reading groups or classrooms. These interactive elements serve as helpful guides through the forest of themes and ideas.

Another key aspect worth mentioning is the frequent use of the pronoun *we*. In the Western world, the ultimate agency on the individual *I/me* makes using *we* a bit more controversial. But there are many layers to this *we* that might be worth sharing. It's both an inclusive gesture, because we're on spaceship Earth together (to borrow a term from R. Buckminster Fuller), but we don't presume that everyone agrees with the ideas, queries, or wonderments presented in the following pages. We also don't assume that identities and positionalities fit so comfortably into one single pronoun, whatever it may be. There are many *we's*, but there's also a sense of our collective presence and impact on the Earth.

From an ecological standpoint, the use of *we* serves herein as an open invitation to recognize what we all have in common as living beings. But it also speaks to our special responsibility as human beings. After all, humans are the cause of global warming and species extinction, not other living beings on the planet like bears or turtles.

Not unlike the ideas presented in this book, the *we* that we're speaking with and about is constantly being reshaped and adapting with exposure to different experiences, ideas, or views, as well as our deeper relationships with more-than-humans. As the book encourages throughout, the use of *we* is meant to be an invitation to hold space for many perspectives and to recognize the other as already part of ourselves. That's the way of becoming ecological – being in multiplicity and possibility as a collective species – which is also to consider "how to inhabit the Earth?"[4]

In this sense, the pronoun *we* is deeply ecological because it's interdependent on other living systems that are constantly changing. All living organisms share existence, and this relationship binds us. So *we* might also include, at times, the nonhuman world or animals and living organisms – more often referred to as *more-than-human* to factor humans as part of this ecological web of relationships rather than separate from it.[5]

Such a foundation is what this book attempts to grow out of. It's a reminder that *we* – as living, ecological beings sharing this planet – are interconnected. As such, the socioecological systems that intertwine must reflect how everyone continues to make meaning together to repair past harms and collectively build generative futures. It's also a spirited exploration of many elements coming together that swirl around our lives and serve to illuminate that, contrary to a lot of popular opinion out there, the collective "we" have significant influence on the future.

As the planet now sits at the crossroads of several possible futures, there's an opportunity to reflect on the power of our language and action in shaping the future of this shared home. This book asks us to reconsider our own contributions to the ongoing narrative of our planet, fostering a deeper appreciation for the symbiotic relationship between language and the environment. Many thanks for joining in this dialogue and for exploring becoming ecological with us.

D.G. and K.J.
Vancouver, Canada

CHAPTER ONE

Discoursing

Words create ecological worlds

Climate change is much more than rising temperatures. It signals unraveling of the very systems that sustain life. The collapsing of ecological systems that are basic to survival has moved to the forefront of public consciousness owing to urgent environmental issues, from extreme weather events and biodiversity loss to resource scarcity and social unrest. Many living organisms, including humans, have already become extinct or are facing it. Despite this, it's still hotly debated.

We're living in what environmental philosopher Timothy Morton and journalist Elizabeth Kolbert, among others, have been calling an *age of mass extinction*.[1] Such an expression invites a more unified and thoughtful way of responding to the world around us, even amid uncertain futures.

For many people around the planet experiencing such rapid changes, this is likely nothing new.

If you're diving into this book, then you probably already have some idea about the challenges the planet faces. To pay attention to any news, or what's right in front of our eyes, is to understand the dire weather changes and the suffering this has brought around the world or even to your local community. Despite such a seemingly dim situation for the future, our intention isn't so much to enumerate increasing catastrophes. We're attempting to offer an unusual blueprint for actionable change.

We concede right away that this isn't your typical ecological book focused on nature, sustainability, or climate, although those topics are certainly present throughout. Instead of looking at *what is ecology*, it dives into *what it means to be ecological*. It's all about *living* with our ecological realities – the lifelong *process* of *becoming ecological*.[2]

Rather than listing facts, stats, or step-by-step guides – or trying to scare people into action through apocalyptic scenarios – this approach focuses on gradual learning and transformation. Becoming ecological has no fixed endpoint and isn't reserved for those who know more about environmental issues than others. It's an ongoing process that everyone experiences and shapes together.

To understand winter, as the poet Wallace Stevens penned, "One must have a mind of winter." In other words, one must *become* winter:

> To regard the frost and the boughs
> Of the pine-trees crusted with snow;

And have been cold a long time
To behold the junipers shagged with ice
The spruces rough in the distance glitter …

The mind here knows winter by *becoming* – listening *in* rather than *to* the snow to experience the interconnected relationship: "For the listener, who listens in the snow."[3] As we show, this is just a brief glimpse of becoming ecological framed poetically through language and meaning.

This difference in approach of *becoming* as opposed to *facting* is rather large. And it can be understood as a contrast between *matters of fact* and *matters of concern*, concepts emphasized by the sociologist and philosopher Bruno Latour, where matters of concern shape the contours of this book.[4] Within the pages that follow, we push to the forefront a concept often overshadowed in conversations about the environment – the profound influence of language and meaning in shaping and then steering the course of ecological transformation.

And yet language may seem like an odd focus for a book about living more attuned with the ecological world. Adding another layer to this, our focus on language is far less traditional than some may expect. In fact, many liberties are taken here, and some might wonder if it's really about language at all. Well, that's part of the appealing ambiguity because language is so much more than it may initially seem.

Far beyond a simple communication tool, or a topic only for linguists, language emerges as a dynamic force, a catalyst that shapes perceptions, constructs meanings, and ultimately molds the understandings that help guide social and ecological change.[5] Language evolves over time not merely by rote learning of words or grammar, but also, as the philosopher of language Charles Taylor identifies, by how narrative and metaphor shape our worlds by bringing in new experiences.[6] Language, to put it more directly, is a conveyor of meaning in everyday life.

Consider this, then, as another way to illustrate. There's a growing emphasis on environmental justice and equity, recognizing (quite rightly) that marginalized communities often bear the brunt of ecological degradation and climate impacts.

Take, for instance, terms like "environmental racism," "climate justice," and "just transition" that are now central to conversations about ensuring fairness and inclusive practices in environmental decision-making and policy implementation. These evolving terms reflect broader cultural and political shifts that shape how we address ecological change. Even as the winds of political leadership shift, the momentum for equitable engagement with the planet persists. Exploring the social and linguistic dimensions of language reveals how specific terminologies and narratives variously influence our daily lives and the perception of being ecological.

Even as these terms have evolved, they may still fail to resonate with broader audiences. A *National Observer* article questioned whether "climate change" is the right term for this current moment, suggesting that framing issues such as "pollution" might be more effective. Why? Because it relates to everyone. Instead of focusing on the extremes of hopeful or fearful messaging, as climate movements have done now for decades, the article highlights how language shapes perception and people's responses. This might involve framing it as "carbon pollution" from fossil fuels rather than "carbon emissions" causing "greenhouse gases."[7]

While not often acknowledged in wider culture, it's true that language contributes to the creation of environmental norms, the establishment of societal values, and the construction of ecological identities. There have always been people in English-speaking contexts who have connected with the land in ways that differ from mainstream Western perspectives. It's through conscious awareness, along with cultural evolution, that environmental discourse has shifted and continues to evolve.

There's more to this. It's important to confess in this opening chapter that we don't just approach communication merely as a way to pass on information. We see language and meaning as powerful forces that continuously shape and reshape cultural and social paradigms. Novelist Ursula K. Le Guin captured this idea in her 2014 National Book Foundation Medal acceptance speech, stating that "any human power can be resisted and changed by human beings." Her words highlight the transformative potential of culture and the "art of words" in driving meaningful change.[8]

In following these invitations, we try to show how language can challenge existing norms, question established hierarchies, and open doors for alternative perspectives on how to value our ecological realities. Language is more than just words, phrases, or a tool for everyday communication. It actively shapes and transforms our perception and worldviews. This is also why language can be divisive and create social conflict. Although this applies to any language, this book will primarily focus on dominant forms of English within the Western context, particularly through mainstream conversations about "environmentalism" as an umbrella concept.

English, as a language, cultural practice, and worldview, presents both challenges and opportunities for enhancing ecological awareness. Language, overall, depends on many types of relationships and requires a collaborative responsibility to create any sort of meaning with it.

The point here is that language doesn't just convey information but actively creates meaning in our own lives and the world around us. Understanding its use and function can change the ways we think, act, and speak.

We're reminded of the comment by Mark Twain about language: "The difference between the *almost right* word and the *right* word is really a large matter – 'tis the difference between the lightning bug and the lightning."[9] Twain's oft-repeated witticism recognizes how language is much more than words. It's also about recognizing that *words create worlds*, changing not only meaning but also perception through meaning.

When evolving and adapting to environmental pressures in a delicate balance with its surroundings, language is just like any other ecological system. What we're calling *discursive ecologies* – the complex and evolving interconnections among ecology, language, and meaning – provides key insights into how better to articulate effective responses to increasing threats to the planet.

The concept of discursive ecologies – essentially,
talking, thinking, and acting ecologically – explores
how the ways we consider and speak about,
engage with, and communicate ecological issues
shape our understanding of and relationship
with the environment.

While this term underpins the overall approach in the book, which is why we'll explain it further in this chapter, we also show in broader ways how it might appear across a range of concepts, themes, and examples in each chapter. By identifying the conditions for *becoming ecological*, which acknowledges that the interdependence of living systems is always in process, we encourage readers to dive into the enduring dialogue that defines our ecological existence as living beings on this planet.

Setting up the dialogue

This introductory chapter is an invitation into a looping conversation between language, action, and ecology. It offers a dynamic interplay where meaning unfolds, much like the collaborative process we experienced as co-authors. Perhaps this approach could be mistaken for sidestepping or even avoiding simplicity. Believe us: if a clear five-step plan existed to help us become ecological through language, we'd gladly adopt it.

But the truth is that neither ecology nor language is easy or straightforward. Both thrive in circular, iterative patterns, with the latter relying on a constant exchange between words and actions. Another way of saying this is that they are systemic, not linear. This cyclical rhythm shapes the book's structure, with chapters designed to reflect this progressive back-and-forth style. Accepting this design as a starting point can be useful, but not essential.

As we've often found, insisting on a linear path in a nonlinear planet can feel frustrating. There's no clear-cut journey from point A (language) to B (action) because the two are interconnected, two sides of the same coin, distinct yet interdependent in how they shape our relationships with existence on this planet.

As the authors – and often imperfect translators of these ideas – we navigate this flow while confronting disjunctions between ourselves, the concepts we engage with, and the real-world contexts we work in. This book is a dialogue on multiple levels: between us as co-authors, among the ideas we explore, and between the nonlinear structure in contrast with the usual expectations of a linear narrative.

The truth is that we're entering and exiting this ongoing collective dialogue in moments of time. This time of *now* will also be changing as these words are being read. So, by committing to talking and acting ecologically, we've embraced this path as a way to highlight often-overlooked aspects of our relationship with ecologies and find our way to greater clarity around achieving objectives of sustainability. Now readers can also be part of this ongoing dialogue, shaping any number of possible futures as collaborators.

Becoming ecological means treating everything we do as part of a *dialogue*, communicating with each other and all living beings through conventional and unconventional forms of language.

Dialogic communication – co-creating meaning with and among others through various forms of spoken and unspoken languages – offers a unifying approach that flows through and connects the systems we are part of. This approach isn't just about how we act but also about how we think and talk about the world.

Building on this, becoming is a dialogical process – an ongoing exchange – where speaking and acting ecologically serve not as rigid prescriptions but as entry points into deeper transformation. To become ecological is to accept our relational natures, where change isn't only about isolated actions or words used by what we might call an individual but more so about the shifts that ripple through entire social or ecological systems.

While speech and action may appear to begin with individuals, true becoming happens when those changes extend outward, affecting the broader web of relationships. In this sense, we're always ecological. But by consciously becoming, we engage more fully in systemic, collective transformation. Talking and acting ecologically, then, are not the end but the beginning of a more holistic shift.

So, as we can see, dialogue is essential. It's fundamental to existence, to living ecologically. This is true not just for human society but also for connecting with and caring for the planet's natural systems, also called the *biosphere*, as a total sum of the planet's ecosystems.

For instance, many Indigenous land practices often consider the more-than-human world as a partner in dialogue, built on respect and reciprocity. In the same way, conversations about the environment could feel like a dialogue *with* the living planet, not something factual or extractive. This kind of relational approach leads to meaningful understanding and action. If living things can't exist without an *environment*, as ecologist Kinji Imanishi made clear, then the same could be said for dialogue.[10] Communication occupies and influences various real and imagined environments.

How we talk about climate change – whether we call it a "disaster" or an "opportunity for change" – greatly impacts how people respond. One path induces fear but also triggers apathy; the other feels more like an invitation to participate collectively.

Even the term "climate deniers" doesn't help move the conversation toward shared meaning. A lot of people who resist climate action don't totally dismiss the value of what they might think of as "nature." These same people may resist terms like "climate action" or "climate justice" because of how they appear to conflict with their own values and worldviews, which are often conservative and therefore focused on preserving the past rather than embracing future change.

As environmental communication researcher Kamyar Razavi points out, using language that ties to traditional Christian values of "purity" and "pristine" nature, such as "pure water" and "clean air," could help bring some groups on board with ecological action, though it might not look or sound the way some people would expect.[11] It might be frustrating for those committed to living more sustainably, but when so many people resist "climate" efforts (even if their reasons may seem illogical), it's worth considering a shift in communication. When becoming ecological is rooted in dialogue, there's an opportunity to embrace different ways of using language to drive necessary change.

Shifting perspectives is a key way of moving forward while also rewriting the past. What if we were to celebrate the inevitable glitches or messes in this process as opportunities, much like how ecosystems adapt to unexpected shifts in their environments? These disruptions reflect the richness of language ecologies, which often defy logical or straightforward thinking. Isn't that freeing?

Instead of prescribing concrete solutions, perhaps a more flexible and organic way of speaking and acting ecologically could be embraced, one that connects diverse experiences and perspectives in a way that encourages meaningful engagement with socioecological issues. That's part of the cyclical nature of ecology and language. Just consider the well-trodden adage "the pen is mightier than the sword." The takeaway is simple: language has immense power to do things.[12]

The *doing* behind meaning complements the *speaking*, and this interplay becomes evident in chapters like Ingesting, Composting, and Unsettling – actions that help us acknowledge how we're always in a process of becoming ecological. Other chapters like Becoming, Storying, and this one,

Discoursing, expand on how language and action are themselves ecological acts. The remainder of this chapter shows how this interaction unfolds, offering a lens to consider discourse as integral to ecology.

How to talk and act ecologically

Many people around the world can feel the effects of the world getting hotter. While many of these people understand the effects of rising temperatures – correlated with ever more present wildfires, food shortages, losses of homes, or rising sea levels – fewer people have developed accessible language to talk about and then act on the realities and fears of ecological collapse with their friends and families. We, too, can fall into this trap.

To add to this need for flexible language, climate and related ecological issues are hotly debated, and misinformation runs amok. As a result, people are often not as practiced with the nuanced language needed to talk about ecological issues and the potential impact these issues may have in our lives. And, in fact, language can often exclude people from pressing conversations about conservation and sustainability.

Beyond developing language that could inclusively communicate or represent the range of environmental and climate-related realities encroaching in our everyday lives, we also need the ability to shift perceptions, which includes attitudes and behaviors, that largely contribute to such realities. One of the many challenges of the all-encompassing term *climate change* is representing its scale and magnitude in a way that's understandable and accessible, especially regarding how it affects humans and nonhumans alike. "Climate change defeats our attempts at perspective," as Dougald Hine puts it in the book *At Work in the Ruins*, because it defies our attempts to talk about it.[13] So we wanted to find a way of addressing this phenomenon through the common meanings people share and ways of living on this planet.

The connection between language and ecology, what we call *discursive ecologies*, attempts to capture how language systems can evolve and positively impact all life on Earth. When people speak of the Anthropocene – a geological era roughly from the late eighteenth century to the present that's heavily impacted by human activities on Earth's ecosystems – they acknowledge the profound global influence of language on human actions and, consequently, on all aspects of existence. Nothing escapes human influence on this planet.

The planet is made of living things that interact with other living things. Humans are only *one* of these living things. So is language. As a living and growing entity, language can help us to become aware of the interconnected worlds we inhabit and rely on for our existence. We can only hope that collective attention to embracing the interplay between language and ecology will lead us to a more prescient awareness of what it means to be ecologically rooted, to live like there *is* a tomorrow. Collectively, we need to shift from total *loss* to more of *less*.

Before moving forward, it's worth pausing to unpack what we mean by discursive ecologies. The phrase may sound abstract, or even too academic, but it actually captures the complexity of the topic quite well. We might then begin by showing a simple example before defining it. Referring to renewable energy as "clean energy" rather than just "alternative energy" is what we might call a *discursive ecological move*.

In this instance, meaning and influence shift with the terminology, while also affecting the social and ecological systems involving energy supply and use. Just look at how "alternative" sounds like something different, a change of pace, though not specific or clearly associated with action. But what "clean" highlights as nonpolluting forms of energy also encourages collective support. After all, does anyone want "filthy" energy?

A shift in language is much more than just word changes, however. Shifting everyday expressions can unknowingly perpetuate broader social meanings in the ways people talk and act around energy, reinforcing power imbalances in energy supply and its ecological consequences.

Discursive, which is the adjective form of the noun *discourse*, describes the features and patterns observed within all the expressions (linguistic and otherwise) about a topic. As shown in the above example of "clean energy," discourse highlights how language systems actively shape our thoughts and actions.

Because discourse isn't a simple concept to wrap our heads around, it might be better defined as the relationships among language, action, and knowledge. Or, put even more simply, *discourse is an interactive way of talking, acting, and thinking*.[14]

Discourse could be considered in two interrelated ways: a) as the languages and practices through which knowledge is shared and power is constructed and sustained in society and b) as the everyday use of language and communication that creates meaning in our lives. When combined, these perspectives reveal how everyone's role in society is influenced by language in daily life.

Take climate change again, an example we use often since it's remained so controversial as a concept: some media frame it as a major crisis we need to act on, while others downplay it or deny it, which can sway public opinion and policies.

Politicians do this too. One person might call it a "global emergency" to push for action, while another might say it's an "economic challenge" to focus on costs. Or even another might call it a "con job" to promote the oil and gas industry. These choices aren't random. They reflect their priorities and how they want to influence people through power and control. It's a reminder that the way we talk about issues matters just as much as the issues or *matterings* themselves.

We're playing with the grammar of this concept even further, as the title of this chapter indicates, by turning discourse into a progressive verb, as *discoursing*, which offers certain degree of poetic license for the book. This layered and densely packed concept recognizes that discourses – encompassing language practices, communication patterns, and knowledge frameworks – play a crucial role in affecting our understanding of ecological matters.

Let's also take a quick look at *ecology* – or the scientific study of the interactions between organisms and their environment, encompassing both living organisms and the material factors that constitute and impact their habitats. People who study ecology seek to understand the relationships between organisms, populations, and communities, in what are called ecosystems, as well as the flow of energy and nutrients through these systems.

Stemming from the Greek word *oikos*, meaning "house" or "environment," ecology is deeply tied to the concept of "home" and "place." As ecological disasters accelerate, they not only physically but also emotionally disrupt people's sense of belonging to home, whether that's our room, house, community, or planet. We all share the responsibility of this *home* planet, as the Chinese poet Li Po observed thousands of years ago: "Heaven and earth – the whole cosmos – is just a guest-house; / it hosts all beings together."[15]

For this reason, we use *ecological* not just as a scientific way to understand the relationships between living organisms but also to highlight the deeply interconnected existence as "beings together" in our shared "guest-house" on this planet. While we use *ecological* broadly, we also use the words *environmental* or *environment* in cases that are explaining surroundings or conditions that encompass natural elements and human-made structures. Put differently, environments are dynamic and symbiotic: shaped by living beings just as much as living beings are shaped by both "natural" and "built" environments.

Ecology can then be viewed as wide-ranging discourse that illuminates how organisms communicate with their environments, how they interact with and act to other organisms, and how these interactions shape the structure and function of ecosystems. The terms *ecology* or *ecological* are *polysemous*, meaning they have multiple related meanings within one word, phrase, or sign. This is like the use of metaphor in language, which always means more than it literally says, like when we sing "you are my sunshine, my only sunshine." Other words that hold various complex meanings include *nature, culture, gender*, and *community*.

Bringing this all together, discursive ecologies refer to the interconnected systems of language, communication, and knowledge that shape how people and communities interact, especially around environmental engagement. While the concept might seem dense, we often simplify it as "talking and acting ecologically" or using terms like "language ecologies."

Let's illustrate this a bit further. Look at the *discursive* example of natural gas, which assumes a fossil fuel that produces a high degree of greenhouse gas emissions is "natural" and a viable alternative to oil and coal. Based on research at Cornell University, "natural" (methane) gas can be as much as 24 percent worse for the climate than coal.[16] And yet it's often viewed and marketed as the best eco-friendly energy solution because of the connotations of the words and the subsequent meaning that has been presented. Let's reframe this meaning: methane gas is a highly polluting fossil fuel that significantly contributes to atmospheric changes. No wonder it's not called "more-polluting gas," even if that's what it is, literally speaking.

Also consider the term *tar sands*, also known as oil sands, to describe the Athabasca River area in northern Alberta in Canada. Tar sands sound dirty, like they need cleaning, much like oil sands do. By making this place sound dirty and unappealing, people would then accept its destruction rather than see it as a unique natural habitat. When confronted with the stark reality of its devastation and the scale of areas and populations being affected, feelings of hopelessness and compassion burnout affect many well-intentioned people. We still need to heat homes, drive to work, and so on. The scale of damage seems to make small gestures like carpooling or shifting to renewable energy sources insignificant by comparison.

Because languages and ecologies are complex, the way to understand their interactive nature can be equally intricate. We believe it's vitally important for lasting societal shifts. We also recognize that using discourse, which is itself a system invented by humans, can create some tensions when trying to build meaningful connections with the more-than-human world. That's why we lean into the concept of discursive ecologies – to highlight how human-centered language can be expanded and reshaped to better reflect the interconnectedness of the ecological systems we are part of.

EXPERIMENTING WITH POSSIBILITIES

Imagine you're walking along a trail and meet someone who claims to be a visitor from a distant future where humans have learned how to be more conscious about *talking, thinking, and acting ecologically*. Curious, you ask them what that even means. They go on to explain that in their time, language has evolved to reflect the deep interconnectedness of all life. Words like "natural" no longer obscure harm, and phrases like "climate change" carry the weight of shared responsibility. Every conversation – whether about energy, food, or relationships – centers on the well-being of ecosystems. Reflecting on this little thought experiment, you might ask yourself: how would your conversations and actions need to change now to create that future? Could the words or dialogue you choose today reshape how others perceive and act on ecological realities tomorrow?

Activating language

To understand how language could be perceived as eco-logical, we can briefly look at literacy through alphabetic knowledge. In alphabetic languages, for instance, words are often used to compartmentalize and categorize natural phenomena, reducing the complexity of ecosystems into isolated objects.

Freezing words in time, which is exactly what written alphabets do, objectifies and devalues the living elements of language. While useful for learning languages, this also impacts how we think about ecologies. When using English words, such as "tree," "river," or "soil," we may overlook the interdependence of these elements within an ecosystem, seeing them only as functional objects to label or identify other particular actions.

What happens to a tree when it's cut down?

The identity of the tree as an object (noun) becomes secondary to the action (verb) of cutting it down. In English, there is a distinct separation between nouns, which represent static things, and verbs, which represent dynamic actions. In other languages, such as Chinese and Japanese, this distinction is often blurred or integrated. Combined verb-nouns in these languages integrate objects with actions, reflecting a worldview where objects aren't isolated entities but integral parts of ongoing processes.[17] If the tree is only an object isolated from everything else, then our perception of that value or meaning would naturally diminish.

Or we can look at how a river's meaning as a verb-noun comes from its relationships with the environment – including flora, fauna, soils, and salmon – forming a dynamic system. The decline of Pacific Northwest salmon exemplifies this ecological and linguistic interdependence. As keystone species, salmon transfer nutrients from oceans to rivers and surrounding forests through bears or eagles, sustaining ecosystems and holding cultural importance for Indigenous communities. Overfishing, habitat loss, and ongoing changes in climate disrupt these relationships, destabilizing ecosystems and straining cultural ties. Here, a river as an object isn't static or separate from its action.

Aldo Leopold, a mid-twentieth century conservationist and author of *A Sand County Almanac*, invited his audiences to consider the concept of *land ethic*, which reimagines humanity's role within the natural world as members of a biotic community rather than its conquerors. Using poetic and metaphoric language, Leopold transforms the seemingly inert noun "soil" into a dynamic, interdependent entity.

Below he likens ecosystems to a symphony, where every organism contributes to the flow of life:

> The land, then, is not merely soil; it is a fountain of energy flowing through a circuit of soils, plants, and animals. Food chains are the living channels which conduct energy upward; death and decay return it to the soil. The circuit is not closed: some energy is dissipated in decay, some is added by absorption from the air, some is stored in soils, peats, and long-lived forests; but it is a sustained circuit, like a slowly augmented revolving fund of life.[18]

Leopold's metaphor of a circuit or symphony invites us to see the land as a living system of interconnection and mutual dependence, urging a shift in perspective that honors the delicate balance of life.

Reducing a river, tree, or soil to mere nouns might seem trivial, prompting some to think, "why are we focused on verbs when the world is burning?" Nouns reinforce a fragmented perspective in our thinking. If we see a river as something separate, as a noun implies, it becomes easier to justify dumping toxic waste into it because it's abstract or distant.

Alphabetic languages, which prioritize linear and abstract thinking as ways of understanding the elements that represent or make meaning, can overshadow more holistic and intuitive ways of connecting with our environments. They can also suppress or silence other forms of intelligence and communication.

Western thinking has emerged from languages influencing English, such as Latin and Greek, where knowledge must be represented by explicit signs or symbols. Language, in this sense, can either enhance or limit our ability to engage with ecological issues in a nuanced, interconnected way, depending on the linguistic frameworks where we are immersed.

Even though language shapes how we understand reality, it isn't limited to speech or alphabets. There are many ways to engage in ecological communication beyond spoken language. *Shinrin-yoku* (森林浴), translated from Japanese as "forest bathing," is a practice that fosters a sense of connection with trees and the world around us, calming our nervous systems and providing relief from the pressures of modern living. Research highlights how this practice strengthens the immune systems' ability to fight infections.[19]

In the past, many people regularly experienced this connection when walking through wooded paths to fetch water, working in orchards, or simply living closer to green spaces. In today's modern world, activities like hiking in a national park, sitting quietly in a city arboretum, or strolling through a community forest have become intentionally scheduled events for many of us, rather than everyday activities.

And yet reducing forest bathing to just a stress-relief exercise misses the broader point. Walking through a forest trail, feeling the crunch of leaves underfoot, hearing an oriole's birdsong, or experiencing the sway of branches in the wind cultivates an immersive experience of *becoming* with the forest. It's an integrated process of engaging with the living world around us, providing a different form of communication outside of conventional language systems.

A simple but effective way to shift our thinking and boost ecological awareness is by using more action-based language, like focusing on verbs. Language assists in relating – to one another, communities, societies, and so forth – by activating or verbing speech as a process rather than outcome. Nouns, as explained further in the chapter Relating, name and give power to an inert or static thing, whereas verbs are assumed to animate or move nouns.

What if nouns could also be animated by reimaging how language and communication functions? Wouldn't this alter the shared meanings and fundamental understandings among people and other living beings?

Marcia Bjornerud writes in her book *Timefulness: How Thinking Like a Geologist Can Save the World*: "one begins to understand that rocks are not nouns but verbs – visible evidence of processes." To understand geology, which is the study of the Earth, one must also think of ecology in terms of verbs. Rocks witness the historic changes over time, making duration visible.[20]

Similarly, the structures of communication and knowledge can represent the invisible elements of social evolution. Words have a history of use, some going back to roots over 1,000 years old. Consider a word we often use in this book: *meaning*. It's usually treated as a noun – something we search for or try to understand. But its roots tell a more dynamic story.

Meaning comes from the Old English verb *to mean*, which itself stems from the ancient Indo-European root *men-*, meaning "to think."[21] This suggests that *meaning* isn't just a static thing to be found but also an action, a process. Meaning is something we *do*, not just something we *have*.

Speaking with the Earth requires a discursive shift in our language, a new way of talking and acting ecologically to highlight meaning. This is true of humans, too, not only rocks. The designer and futurist Buckminster Fuller created a book of images, writings, and design principles titled *I Seem to Be a Verb*, where he acknowledged how even humans as active beings are more verbs than nouns because of our capacity for constant movement and adaptation:

> I live on Earth at present,
> And I don't know what I am.
> I know that I am not a category.
> I am not a thing – a noun.
> I seem to be a verb,
> And evolutionary process –
> An integral function of the universe.[22]

While not as dominant in our knowledge and communication systems, these models of treating the world as active, not static, do exist across many different cultures.

As verbs, we're integral to the universe and animated through our grammar of discursive ecologies. Verbs are inherently more ecological because they embody constant exchange of energy through action, movement, and interaction, all of which reflect the dynamic and interconnected nature of ecological systems.

One of the key aspects of ecology is *diversity* – cultivating a wide variety of species, biota, flora, and organisms in a social or biological system. Just as diverse ecosystems flourish while monocultures wither, societies also thrive when they embrace diversity, both culturally and ecologically. This is also true linguistically as well. Without conscious invigoration through language and communication, considering their diverse elements, we risk living out of balance with Earth's ecology, jeopardizing our survival and accelerating the already quickened ecosystem collapse leading to mass extinction.

Though not inherently relational like some languages that emphasize interconnectedness, English still provides ways to engage with ecological ideas and promote environmental awareness. Let's look at the fun interplay of "carpool" and "carpooling." In linguistics, this is called a *collocation*, two or more words combining to make a new metaphoric meaning. The word "car" is used as a noun, a thing, but the word "pool" isn't. It's used as a verb, an action of collecting together. When we carpool, we obviously don't go swimming in the back of a truck. We pool our transportation resources.

These kinds of discursive practices are more than mere parlor tricks. They're important because English is currently the most spoken and influential language in the world. We're not calling for a complete overhaul of language but rather a shift in how we view its everyday use – through words, ideas, engagement, and speech – and how language is both shaped by and shapes our perceptions and actions within the socio-ecological systems we're part of and make meaning with.

PRACTICING

Begin by reflecting on how words like *tree*, *river*, or *soil* might feel static or isolated, then contrast this with languages where nouns and verbs blend, emphasizing relationships and processes. By yourself or with others, turn ecosystem elements into verbs. Imagine a *tree* "treeing" (providing shade, nurturing soil) or a *river* "rivering" (nourishing landscapes, carrying life). Take this perspective outdoors or simulate it indoors by observing nature and narrating dynamic verb-centered descriptions of what you see. In a group, consider having a dialogue about how this shift in language deepens ecological understanding and might inspire action. Continue to practice "verbing" environments in your daily life to develop greater awareness of these interconnections.

Living systems of language

Healthy living systems whether linguistic, ecological, or otherwise depend on mutual coexistence, enabling resilience and adaptation. Oddities and irregularities make the system more robust and are fundamental to helping discursive ecologies adapt to changing circumstances. History continues to show that attempts to control or homogenize discourses as living systems of language and meaning often lead to suppression and conflict, as no single expressive repertoire or disciplinary vocabulary can fully address the complexity of living systems.

One prime example of this is the historical suppression of Indigenous languages and cultures around the world during many phases of colonial expansion, which not only erased diverse perspectives but also sparked resistance and long-term cultural losses that are still apparent today. Just as ecosystems collapse when biodiversity is lost, discourse loses its vitality and adaptability when reduced to a single dominant narrative. This could look like silencing disagreement or shutting down different perspectives.

The rapid collapse of ecosystems – evident in the 73 percent decline of species over the past 50 years – is more than an ecological crisis.[23] It's the unraveling of the interconnected systems that sustain life itself. These outcomes reveal the fragility of systems built on extraction and exploitation, urging us to reimagine our relationship with the planet and each other. To do so, we must also rethink the language we use and how we communicate.

Shifting from fear-driven terms like "climate emergency" or "climate crisis" to descriptive concepts like "mass extinction" or "collapse" can maintain urgency while refocusing attention on systemic interdependence rather than immediate alarm. Mass extinction can be understood as a transition rather than just an end, part of Earth systems,[24] much like how the extinction of dinosaurs paved the way for mammals to thrive or how controlled burns in traditional forestry practices cleared the way for new growth and regeneration. Language, too, goes through cycles of transition and reinvention. This reframing invites collective action and emphasizes shared responsibility, implicating everyone in the need for long-term solutions.

If we don't change the language we use to talk and think about the future, we will continue to think old thoughts and come up with outdated responses to these new and unprecedented conditions. As Albert Einstein is credited with saying, "problems will not be solved by the same thinking that created them," where thinking is intertwined with talking and acting.

> Changing language *changes* dominant discourses, which in turn *changes* thinking, and this makes large-scale *change* in how we engage with ecologies possible.

If we treat language and ecology as disconnected, we neglect our best efforts to achieve large-scale, socially mobilized solutions. Systems biologists like Chileans Humberto Maturana and Francisco Varela emphasized that language isn't static, but a dynamic, living process that shapes how we adapt and coevolve within interconnected systems. Language, they argue, is central to how we create meaning and navigate change.[25] The internet, health care, artificial intelligence, or politics, as well as ecosystems, are recognizable examples.

In her roundly praised book *Finding the Mother Tree*, our University of British Columbia colleague Suzanne Simard illustrates the complex communication networks of forests, where trees adapt, collaborate, and respond to each other through chemical signals and electrical impulses.[26] Human language systems are similar: they serve as the foundation for collaboration and innovation, helping us adapt to shifting environmental and social conditions.

One of the most significant insights from living and thinking in systems is the concept of change as a fundamental component of survival. Change isn't just inevitable; it's essential for adaptation. This idea applies not only to ecosystems, which recover and reorganize after disturbances like forest fires, but also to social and linguistic systems. The growing adoption of regenerative practices, such as permaculture and circular economies, reflects this shift in thinking. These practices embody a transition from extractive to restorative systems, demonstrating how new narratives can lead to more sustainable ways of living.

For these shifts to take root, we must reconsider the language we use to frame environmental and social challenges. Fear-inducing and confrontational terms often polarize and alienate, hindering collective engagement. Language could instead promote collaboration and a sense of shared purpose. The term "climate breakdown," for instance, could highlight the disruption of systems while also pointing to opportunities for reorganization and co-creative transitions like in "climate collective." Even amid crisis, there is space for reorganization and for innovation and growth.[27]

This reframing is critical as we navigate what the French complexity thinker Edgar Morin called the *polycrisis*: a time in history characterized by multiple complex global challenges happening at once.[28] Terms like *polycrisis* and *metacrisis*, which are increasingly being used, encompass challenges such as violent conflicts between nations, mass migration due to uninhabitable conditions, and global food shortages triggered by extreme weather events like droughts or floods, all of which are happening simultaneously and are interconnected. By understanding the systemic nature of these challenges, we can move beyond fragmented solutions and focus on holistic approaches.

Language, as a living system, plays a vital role in this process because it's one of humanity's richest resources for stabilizing the future. It has the power to nurture diverse discourses, holding multiple perspectives and adapting our thinking to new realities. By embracing a more inclusive, dynamic, and systems-oriented approach to language, there's greater possibility to build the foundation for a more resilient and equitable future, where human and ecological systems thrive together.

Inviting co-creation

Bringing this chapter to a close, hopefully we can now acknowledge the language we use to address ecological issues is of momentous concern. Collective efforts, whether to responsibly reduce waste, drive less, limit meat consumption, have fewer children, and so on, have not been enough. With many natural systems collapsing, that need for change is upon us, whether we like it or not. Understanding what we can do to shape the future with language gives us a sense of optimism. By transforming everyday discourses, we change the dominant discourses that can shift the future.

In this book, we're diving into discursive ecologies across various contexts, from how we talk about the environment to how culture shapes our ecological views. We see these as interconnected systems where language and power work together to shape meaning in our worlds.

By providing chapters as a series of progressive verbs as active agents for change, this book's structure and content empower readers to grasp the nuances of social dynamics and to actively participate in transforming them. Considering progress and duration, chapters contextualize the ongoing and continuous connections of ecological actions, such as Composting, Unsettling, and Futuring. Additionally, they explore the social and cultural connective tissues that are often marginalized in ecological discussions yet integral to environmental communication, such as Listening, Relating, Witnessing, and Storying.

These chapters serve multiple purposes, as invitations, inquiries, curious tangents and examples, arguments, affirmations, and possibilities. Chapter constructions prioritize circularity over linearity, embracing overlapping and evolving concepts. As such, chapters can be read out of order, just like the many paths to take when walking in the woods.

Readers are encouraged to embark on an adventure with the conceptual and systemic framework of the chapters, fostering curiosity to navigate back and forth, side to side, and up and down, or even setting some aside for later. It's an invitation for open and active engagement over time as circumstances change.

The overarching claim introduced here in the opening chapter, and put forth in each following chapter, is that engaging in discourse is inherently intertwined with ecological considerations. These insights can help cultivate a deeper relationship with social and ecological responsibility, providing guidance on how to enact it. This creative potential of environmental communication and knowledge, grounded in lived experiences, can be a catalyst for generative societal change. At its core, we're always in a process of becoming ecological through the ways we think, act, and talk about the environment.

So then, what's next? Or perhaps it would be better to inquire with a good dose of curiosity: what's does this moment call for? What we can co-create together through this ongoing dialogue is what we propose in the rest of this book.

REFLECTING

- How might the language you use or encounter in conversation about environmental issues shape your understanding of and relationship to the ecological world?
- What are a couple of examples that show the role language plays in shaping our collective understanding and response to ecological challenges, and how can we reshape it to bring forth effective action for the environment?
- How, then, can rethinking and reframing the language we use around ecological and social challenges help build collective engagement that will lead to systemic changes?

CHAPTER TWO

Becoming

Constantly arriving, day after day

If we can accept for a moment that language has the power to shape our perceptions and worldviews, as outlined in the previous chapter, then it inevitably draws us into the thorny issues of reality itself. Historically, this has led to a recurring, profound inquiry: what is the meaning of our existence?

At its core, this question reflects humanity's ongoing search for fulfillment and, in more contemporary terms, achieving "the good life." But what's the motivation and process for achieving it? When we uncover some sense of meaning in our existence, how do we interpret it, understand it, or ultimately apply it to our lives?

These are timeless questions many people have wrestled with for millennia. While definitive answers remain elusive, these inquiries persistently surface. They nag at us in our most honest, reflective moments, resonating across generations and cultures.

Despite these efforts, even the purpose of our existence remains a perpetual enigma veiled in uncertainty and ambiguity. The fear of death often arises not so much from the prospect of dying itself, but more from the concern that our lives and evolving relationships may be cut short, and that we may not have lived life to the fullest. Such an existential worry invites many of us to consider how we can live in a state of wonder and awe as opposed to feeling the crushing anxiety of uncertainty or the universal weight of nonexistence.

After all, isn't the term *climate anxiety* just the fear of not existing someday? And isn't this just a fear of death for ourselves or others arriving sooner than we may want? Purposeful in the way it names fear, the term *climate anxiety* has a specific focus about ecological collapse, and particularly the threat of experiencing this during our lifetimes. The anxiety is about the end of our predictable knowing of what we have come to understand about existence.

We can lighten this up with a piece of good news: becoming offers of way of reducing anxiety, whether that's about the future of the planet or specific weather events. These are just some of the everyday messes of living in today's global reality. But all of them arise within the ongoing process of becoming the ecological beings we already are.[1]

We've all likely heard the common statement: *people don't change*. Well, we don't see it this way. It's quite the opposite, actually, because people are always changing and adapting to the environments around them. In fact, it's the one constant certainty in life: we change, everything changes. This is true psychologically, biophysically, emotionally, and so forth.

> Every living thing is always becoming – continually
> in the present while also changing dynamically
> over time.

Much like the changing seasons – from autumn to winter, spring to summer – they transform while remaining seasons, cycling in and out of existence. But fear of the unknown that leads to various forms of anxiety can get in our way and make change, even necessary change, seem undesirable or impossible.

This chapter explores *becoming* as a continuous process of learning, unlearning, adapting, evolving, and embracing our ecological natures. And language plays a crucial role in shaping this process, influencing how we make sense of and navigate change. As we see with the evolution of language and human development, the conditions that we find ourselves in largely create a sense of who we are and how we might behave in given circumstances.

Our worldviews are contingent upon how our languages and environments are always constructing how and what we think about, as well as our perceptions of reality in our day-to-day existence. While this might seem too philosophical or abstract at times, it bears directly on our daily activities of communication and action. Much like building a regenerative garden, enriched soil takes time to cultivate. Once the soil is nurtured, the garden will yield growth.

The concept of *becoming* that's explored in this chapter invites the possibility of witnessing the change already in process, while it also allows for constant state of adaptation and movement. As with ecological systems, we're all in a state of continuous flux. The sooner we can accept this reality, which includes the anxieties related to any number of possible ecological or social futures, the more we can integrate with the flow of living life rather than resisting it.

To understand becoming, we need to engage, at least partially, with the metaphysical concept of *being*, which shapes how we make sense of existence, and what philosophers call our ontological realities. For example, some cultures see land as a living entity to be respected, while others treat it as a commodity to be bought and sold. These differing ontologies, or worldviews, shape how we interact with society and the environment, as seen in the clash between Western extractive industries and Indigenous and other relational practices seen around the world across various cultures that treat humans as part of and interconnected with nature. As ecological beings, we're always in a state of becoming, whether we consciously recognize it or not.

At this point, the concept of becoming may still be somewhat confusing because we aren't becoming in a sense of *we're not there yet*. We are always becoming because *we're constantly arriving*: day after day, like waking to a new morning, again and again, changing all the while. We exist only in the present moment, yet we're never fixed but always in motion, never the same from one moment to the next.

Our language ecologies are also constantly evolving, shaped by the societal and environmental conditions that influence how we learn and make meaning in times of ecological change. Language is in a process of becoming, where ambiguity and contradiction arise from attempts at communicating meaning among diverse ideas and situations.

If we're honest with ourselves, we quickly realize that we don't keep the same language throughout our lives. Do we have the same vocabulary, expressions, and patterns of speech at ages sixteen and forty-five or even later at eighty-five, should we get there? All people don't ingest the same cultural content of social media, TV, film, and so on. Many parents don't speak to their children in the same way they spoke to their own parents. Even large language models like ChatGPT don't remain the same.

These natural language shifts occur frequently. They must. This is how energy moves through social systems over time. These are the dynamic structures in which language and meaning are embedded. They're always expanding or contracting depending upon the various communicative needs and social environments we inhabit. What we hope to show in the following pages is how this ongoing process of language is one of *becoming* – a dynamic interplay of experience and meaning that extends into daily practices of striving without fully arriving.

Change isn't easy

Accepting the process of change, whether it's changing our opinion, a relationship, a hair style, or even something larger like the direction of a government, isn't an easy path. It can be obvious to see but difficult to shift. This is why many thinkers, poets, philosophers, teachers, and spiritual seekers have attempted to answer the mystifying questions of who we are (identity), what is the nature of the world and the meaning of our existence within it (ontology), how we know what we know (epistemology), and how we should behave (ethics)?

Narrowing these down to an exploration of becoming as it pertains to our ecological life poses both challenges and opportunities. Even if there are more speculations than answers, it's important to make home in uncertainty, a process that enables emerging possibilities to appear.

In the traditional Western perspective, in both the past and the present, there's a common belief in fixed and inherent qualities that define us as *human beings*, like Socrates's *soul*, Aristotle's and Augustine's ideas about an individualized *will*, Freud's *ego*, or the existentialist concern of the *self* as a guide to personal *help* or *well-being*. Some other perspectives consider being as not only the fixed *state* of existing but also the ongoing *process* of existing. This chapter focuses on the latter viewpoint.

The concept of being assumes that our essence or existence isn't fixed; it's a dynamic and changing process where meaning arises out of socioecological existence. All that which *is* in reality can be determined through our *being*, which is a process of changing and passing away according to the Japanese philosopher Kitarō Nishida.[2]

We shape our social meanings just as we shape our relationships with the ecological world, continuously constructing through our lived experiences. Just as we create our own interpretations of events and situations in our social interactions, through stories, languages, perceptions, and so forth, we also construct our perceptions and attitudes toward nature and the environment. Living beings are part of the world and, through this dynamic process, bring meaning to it. This process even appears in everyday language. When we say a river is "angry" or a forest is "healing," we're not just describing a scene but also assigning emotion, agency, and relational meaning to the natural world.

The Confucian viewpoint embedded in Chinese philosophy sees a person more as a "human becoming" than a fixed "human being."[3] In this perspective, being a person is seen as a process of becoming that we engage in through our relationships, rather than something we inherently possess. Personhood, an expression we'll get to later, is about the possibilities that emerge through our interactions.

Consider how we're not simply individuals within a community, but how we become individuals through our relationships in that community, particularly with relationships that are always changing. The psychologist Kenneth Gergen came up with the useful term *multi-beings* to "become together," with capabilities of being many persons, combining multiple experiences and sensations collectively.[4] Similarly, the Vietnamese teacher of peace Thích Nhất Hạnh coined the term *interbeing* to describe the deep interconnection of all life.[5] Across cultures, language has tried to capture the experiences of being and becoming, not as isolated states or identities but as part of our shared ecological nature.

If we think about it, we might see how people develop their identities through connections – other people, communication, beliefs, and even more-than-human entities like animals and trees – rather than as isolated units navigating the world alone. For example, a farmer's identity might be shaped not only by their work but also by their relationship with the land and the community they serve. These relationships are built on the communicative systems that weave together the fabric of society.

While *being* shapes our sense of existence as a whole, *becoming* reflects the ongoing process of change day by day, minute by minute – by adapting, unlearning ingrained habits, and embracing transformation. It resembles an ecosystem, both steady and always changing. Becoming emphasizes fluidity and growth over fixed states, striking a balance between continuity and change.

Becoming is often considered by many as striving toward a future that's disconnected from the present or as trying to become something or someone that doesn't yet exist. We use the term a bit differently. Rather than departing from the present, becoming is about fully inhabiting each moment as it unfolds. Brazilian educator Paulo Freire described this as "beings in the process of becoming," uncompleted and continuously evolving within an unfinished reality.[6] This is also how language works, constantly flowing between people and connecting everyone in collective processes of becoming through discourse.

The ways to see and embrace becoming are displayed in the coming chapters, particularly through practices and perspectives of listening, relating, ingesting, witnessing, storying, and so forth. They encompass where we are now as well as the emergent possibilities of who we might *become*: across our pasts, presents, and futures as one circular experience.

In truth, everyone is always in a state of becoming, whether consciously or not. Becoming is also akin to living.

To be alive is to become, to participate in this world. The practice, then, is to develop awareness of our becoming, rather than striving to get somewhere or be someone other than the person we experience being in each moment.

Being and becoming are intrinsically connected, both highlighting our ecological nature through the interplay of existence and transformation. A tree standing in a forest is an example of *being*, existing as a tree and providing oxygen, shade, soil health, and habitat for various animals, insects, and other organisms, with its role and function seemingly established. This tree is also in a state of *becoming* because it undergoes significant changes over time, growing and evolving each day through shifts in weather and seasons or external factors like insects, animals, or humans. Both highlight transformation, yet they differ in that being appears more constant, while becoming is in continual change. Put another way, the tree is constantly *treeing* while in the process of being a tree.

Now turning to Donna Haraway, as a philosopher of science, consciousness, and technology, can help us deepen our understanding of becoming as a way of communicating and acting ecologically. She writes: "If we appreciate the foolishness of human exceptionalism, then we know that becoming is always becoming *with* – in a contact zone where the outcome, where who is in the world, is at stake."[7] Saying this in slightly different words, our existence is never independent but always entangled with other beings and forces, shaping and being shaped by the worlds we inhabit.

Becoming-with describes how different beings, such as humans, animals, plants, and even technologies, are interconnected and constantly shaping one another. By challenging Western ideas of individualistic supremacy, it shows that beings are in a continuous state of becoming through their connections, not as fixed, separate entities. The addition of the preposition "with" to "becoming" highlights the relational nature of existence, reinforcing that our identities and experiences are shaped through our interactions with others.

This all implicates that our becoming is never isolated. The preposition *with* is always there in ecological language, even if it's not written or spoken. It unfolds alongside other entities within ecological systems, and we can reflect this understanding in our language. Consider how the relationships among bees, flowers, and humans shapes agricultural practices and how we talk about food production, emphasizing mutual dependence rather than separate, independent actions.

——— ∘∘○∘∘ ———

Another way of showing becoming-with could be seen in the widely relatable affiliation between humans and dogs. In this mutually transformative relationship, humans and animals engage in a process of becoming, particularly through forms of verbal and nonverbal communication. The human and the dog shape each other's lives, with humans learning to understand canine behaviors and emotions, while dogs communicate with the habits and emotional cues of humans. Even without the use of spoken language, this relationship constructs a language ecology where the processes of becoming feed back into each other. Over time, as anyone who has lived with animals would know, both partners undergo significant change, cultivating important qualities like empathy, responsibility, and interconnectedness. This example may seem simple and obvious, but it's universal. Beneath our emotional bond with domestic animals lies the agency and subjectivity of becoming-with the more-than-human beings that shape our interconnected world.

In the world of language, there's a fascinating question about how people evolve. This is also called growing and adapting. Rather than being stuck and finding a solution or arriving at a specific ending point, it's more like a constant process of becoming. That is, becoming might always be considered as *with* other living beings on the planet. This involves practices that help us unlearn and unsettle certain ways of being, such as moving on from older habits and values or shifting our language use. We could also look at it like the dynamic actions of organisms in systems that are adjusting to changes and adapting to the interrelated webs of existence.

EXPERIMENTING WITH POSSIBILITIES

Imagine waking up in a world where language never changes, where words remain frozen in time, unable to adapt to new technologies or cultural shifts. Communication would feel rigid, unable to capture evolving meanings or emerging realities. Now contrast this with how language actually works: new words emerge, meanings shift, and the way we express ourselves transforms alongside societal and ecological changes. If we recognize that language itself is always in flux, how might this awareness help us speak, listen, and engage in ways that better reflect an interconnected and evolving world? If language shapes our reality, what happens when we let go of familiar language structures and embrace new ones? How might this process of unlearning and relearning affect who we're becoming?

Making meaning together

Hopefully, now we might have a clearer understanding of becoming. But how can we engage with language and meaning that supports a way of living that's constantly in flow, mirroring ecology, always changing and adapting to the contexts around us?

How can we talk about the realities of living amid ecological collapse while also supporting our growing awareness of ourselves as ecological beings?

A commonly suggested approach is through rational discussion, where individuals argue for or against a topic with an expectation that their perspectives will be considered fairly and evaluated for a logical conclusion. This approach may sometimes lead to agreement, but it fails to drive the deeper social changes needed for systemic transformation because it reinforces siloed perspectives instead. If this dynamic worked, then wouldn't a society presumably steeped in rationality and reason already be less combative?

Another possible path is shifting how we communicate: not talking *to* or *at* one another but *with* one another. The notion of *dialogue*, as considered by the physicist and philosopher David Bohm, among others, provides a concrete practice that forges deeper socioecological connections and opens possibilities for becoming both *with* and *through* our relationships. Through *dialoguing*, also known as *dialogic communication*, we enter relationships that go far beyond the realm of human relationships.

Unlike the word *discussion*, which shares the same origin as the words *percussion* and *concussion*, meaning "to break apart," the term *dialogue* originates from the Greek word *dialogos*, meaning "through or across words." Here, *logos* refers not only to "the word" but also to the broader concept of "meaning." *Dia*, by comparison, means "through," rather than the commonly misperceived "two." According to Bohm, dialogue is defined as the "stream of meaning flowing among and through us and between us."[8]

In dialogue, as opposed to discussion, the collaborative potential of the group or collective promotes the emergence of new meanings and understandings. Perhaps just think of dialogue as an open invitation to actively join in and build meaning together using language, as opposed to the individualistic desire to be seen and heard at the expense of building connection with other living beings.

In education, dialogue, also known as dialogic practice or open dialogue, is helpful because it encourages people to learn by creating meaning together using language in context.[9] Rather than speaking to be heard – "people just don't understand me!" – this view of language emphasizes its collective nature. Language becomes a tool for shared meaning-making, not opposition. Dialogue, then, is a co-creative activity that helps us learn from and with each other as an everyday practice of becoming.

Because dialogue asks us to hold space for multiple perspectives, it can be especially challenging in today's society. As educators, we're familiar with the constraints on dialogue and how to steer conversation away from combative viewpoints that could promote harm. When those interactions enter the dialogical system for prologued periods, they become toxic to the language systems. But such ecologies have the capacity to reconstruct meanings and, in the process, reintegrate conflicting viewpoints as integral players of a communal becoming with each other.

Because dialogue is about connection and creating shared meaning, it differs both in kind and practice from rational arguments that focus on individual opinions and separate experiences, often driving people apart. In today's world, attention often centers on individual interests, where people act based on what benefits them most. Social media debates might focus on asserting personal opinions or attacking individuals rather than seeking common ground. Clickbait headlines like "Famous Podcaster *Destroys* Opponent" or "Conservative Senator *Owns* Liberal" exemplify this divisive dynamic.

In these confrontational media and political landscapes, it may seem unfashionable to advocate for forms of communication that focus on becoming together, rather than pointing fingers and assigning blame. If we can pause our usual assumptions and opinions, dialogue allows the exchange of ideas to shape collective understanding by learning and unlearning through language.

Rather than sticking to one fixed belief, dialogue offers a more expansive way of examining knowledge and experience. The Austrian philosopher of language, Ludwig Wittgenstein, once warned that meaning isn't just something subjective or internal that one person understands – what he called one's *private understanding*. There's a fundamental issue with the assumption that we create meaning on our own. As Wittgenstein suggests, connection isn't made by grasping another person's private meaning through a mental process. Instead, it emerges out of coordination with others, sharing thought and experience together in tandem.[10]

Even when people are alone, we're still in a kind of dialogue, drawing on past experiences or conversations and carrying them into the present while imagining the future. When we experience the aesthetic awe of nature – like pink light illuminating scattered summer clouds – we often already have language to describe it. These shared understandings and collective relationships are what shape any number of possible realities. The words we use to explain a sunrise or autumn leaves changing colors, whether scientific, poetic, descriptive, or emotional, reflect not just individual perception. They also show a larger cultural and ecological context that has influenced how we see and communicate with the world around us through dialogue. Through language, we resonate with our environment.

Let's root this in an example of a community gathering to discuss the impact of plastic pollution on local waterways. During the dialogue, people could share personal anecdotes about witnessing plastic litter in nearby rivers and lakes, recounting experiences of seeing wildlife harmed by plastic debris. These stories could evoke emotions and empathy among the local residents, highlighting the human dimension of the environmental challenges.

As the dialogue progresses, community members could also share success stories of collective efforts to clean up plastic pollution or reduce plastic usage. These stories of action and resilience have the potential to inspire deeper change and motivate further engagement in environmental conservation efforts. Through storytelling, the residents not only deepen their understanding of the issue but also feel empowered to take consequential action to address it, building a sense of collective responsibility and solidarity in the community.

To be clear, though, dialogue isn't just talking to understand another person or to get your own point across and *be seen* or *heard*. It's a contributive effort. Such a shift in individual versus shared understanding may seem counterintuitive. And yet it alleviates tensions among the divided groups and people who are angling to be understood or heard. The process reconsiders how dialogue is a practice of co-creating futures since the energy we put into the discourse now will pass through that system of communication for years to come.

PRACTICING

Set up a *dialogue circle* with some other participants to explore any ecological topic like renewable energy or plastic pollution. Begin by distributing "perspective cards" that represent different viewpoints, such as a farmer, scientist, community activist, or marine animal like an octopus. Each participant speaks from the perspective of their assigned card, describing how the issue affects them and their priorities. After initial sharing, invite the group to move into an open dialogue, shifting from representing individual perspectives to collectively imagining other possibilities. Try to reflect on how language and collaboration shaped your experience. This practicing helps to apply what it's like to step into unfamiliar worldviews while co-creating shared meaning through dialogue.

Reverbing meaning

Everyone's interconnectedness with the world might also be explained through sound waves. In fact, we, the authors, are living proof of this. As two sound artists who embrace improvisation, we experience sound as a language that reflects the process of becoming. Our discovery of reverb as part of language ecology emerged from playing music and listening to the waves, revealing how sound carries relationships and interactions as the echoes of the environments we inhabit.

> Discourses, like ecologies, reverberate across cultures and histories, constructing our experiences of the world.

To gain a deeper understanding of interconnectedness from a different perspective, we can explore how waves illustrate the relationship between language and ecology. Waves, with their rhythmic patterns and interconnected movements, provide a unique lens through which everyone can better grasp the intricate web of the world around us. Can you imagine the constant ebb and flow of waves as a representation of becoming? This perspective highlights the dynamic interplay between opposing forces like permanence and change. Waves, with their rhythmic dance, beautifully capture the essence of this ever-shifting balance and interconnection of time, particularly through materiality and perception.

Reverb, short for *reverberation*, is a phenomenon that occurs when sound waves bounce off surfaces creating a persistence of sound. With each bounce, the sound gets quieter, some of the wave is absorbed. The sound's decay tells us about the enclosed space we're in, whether the surfaces are hard and flat or plush and curved. When you clap your hands in a large empty room, you'll hear the initial clap. This would then be followed by a series of quieter repetitions as the sound waves reflect off walls, ceilings, furniture, and other surfaces. This creates a sense of space and dimension in the sound. It also highlights human influence on and by ecological factors, where reverberations of sounds, as much as words, actions, and behaviors, impact the environment and everything becoming within it.

Consider how different environments and materials, or even language and communication, can shape the reverberation characteristics, impacting not only the sound in recordings or live performances but also the overall ambiance or mood in various spaces. The choice of surroundings and materials can significantly influence the auditory experience, adding a unique flavor to the soundscape. Likewise, the choice of words, expressions, or tone can impact our social experiences.

Reverb, then, can be likened to the interconnected reverberations within an ecosystem. Like sound waves reflecting off surfaces in a room, ecological reverberation occurs when interactions and influences persist in an environment after the initial stimulus has ceased. This is also called an *echo effect*. Try to imagine an isolated ecosystem, like a remote forested island, where every sound or movement has settled into a delicate rhythm over time.

Now introduce a new species. Perhaps this is an invasive bird with a piercing call that disrupts the usual soundscape. The arrival of this new presence is like a sudden loud clap in a small room, where the echoes bounce unpredictably, altering the space. Other species must now adjust, whether by shifting their calls, changing their feeding patterns, or even relocating. Just as sound waves reverberate and reshape a room's acoustics, ecological relationships shift and adapt in response to new disturbances, illustrating the ongoing, dynamic process of becoming.

As we can see, the effect of reverberating extends beyond sound to illustrate the enduring effects and interconnected nature of processes within ecological systems over different durations. On a grand scale of time in the modern period, over the past few hundred years, we're now witnessing this percussive feedback from the impacts of human populations on all aspects of the environment.

To illustrate, let's consider what we might call as a short-hand *ecological reverbing* in the next few instances:

- When large areas of forests are cleared for agriculture, urban development, or logging, it not only directly impacts the species living within that area but also reverberates throughout the entire ecosystem. Cutting down trees reverberates beyond removing the forest, disrupting wildlife habitats and migration patterns, water cycles, and even the local climate, as well as accelerating soil erosion and floods and droughts.
- In the case of whale populations, there are also literal problems from reverberation. Sound waves can travel further in water than air because the medium is denser and more consistent. Whales communicate over great distances with their pods. Ocean traffic and industry produce very loud vibrations that make it impossible for whales to hear each other, and this is an important factor in their decline.
- Another example of ecological reverbing is microwaves from communications devices and cell towers causing "sudden hive death" among bees, often thought of as a key species in the interdependence of pollinators and global food supplies. Deafened by this noise that's imperceptible to humans, bees can no longer find their way in the world or back to their hives.

Just as reverb in sound reflects off various surfaces, language reverberates through human interactions, influencing and connecting individuals and communities. The ways we communicate, share ideas, and express ourselves create a web of connections, our mutual becoming within society, mirroring interconnected ecosystems. In other words, language exhibits elements of reverb as a dimension of ecology.

Reverberating communication amplifies and spreads sound, quickly filling any silence and leaving little space for quiet reflection or balance. As public discourse becomes more crowded with media that amplifies opinions and digital noise, politicians or leaders often raise their volume to capture attention and shape the conversation, relying on the echo effect across society. Sound familiar?

Social media also serves as a prime location of the reverb in language ecologies. When we send reply to a message or make a post, it reverberates off social groups. It can either enhance a conversation, building connection, or perpetuate disconnection with certain subgroups. The impact of our messages extends beyond the words themselves, influencing the dynamics of our social systems through these interactions.

Regardless of these invasive examples, it's encouraging that language can also spark ripple effects that lead to changes in behavior and beliefs across society. The words we use, the stories we create, or the messages we convey can persist in the collective memory of a society, reverbing across many generations. This linguistic reverb shapes cultural norms while also influencing attitudes. These effects on the ongoing dynamics of human communities are similar to how echoes linger in a physical space, but they decay over longer durations.

When a new phrase or concept emerges, it spreads and evolves, shaping how people think and interact. Much like how ecosystems adapt to environmental changes, language evolves through new ideas. Consider as an example how *well-being* started as a basic health concept but has grown into a widespread practice that reshapes how we think about physical, mental, and spiritual health, just as species adapt to survive in shifting habitats.

Reverberating, or reverbing, like many other verbs used in this book, contains multiple layers of meaning. To use language is also to *re-verb*, prompting us to reconsider how we apply language for societal and ecological change. Becoming highlights the reverberation of constant change that serves as a key concept for recognizing our ecological capacities as living beings on this planet.

Ecological remix

What if someone you know starts adopting eco-friendly practices in their daily life, like reducing single-use plastic, recycling, composting, or prioritizing sustainable products? This person is becoming ecological by actively incorporating environmentally conscious habits.

But the experience of becoming goes well beyond these simple actions. While these everyday habits are important, they're surface-level changes. Becoming involves a more profound transformation of assumptions about reality. It also asks us to look at core beliefs, questioning the way to see the world beyond our individual "self," and then integrating these beliefs as practices through language and action.

Drawing on some of the ideas in this chapter, we might view becoming ecological like an *ecological remix* – the ongoing, adaptive process of attuning to our surroundings. Similar to remixing a song, it involves weaving together past knowledge, present experience, and emerging rhythms to stay in relationship with a changing world. It's a collective process, one that relies on resonating with multiple elements, including other people, groups, perspectives, cultures, or what we might more broadly call ecologies.

This process of becoming is what in improvisational circles would be called a "yes, and …" approach, where participants build on each other's ideas by accepting what was said ("yes") and adding their own thoughts or extensions ("and …"). Such an exercise encourages constructing meaning together, remixing and layering various perspectives together rather than replacing a perspective with another one.

This chapter echoes the previous one in exploring how language shapes our ecological worldviews and behaviors because becoming and discoursing are the same process. More specifically, it offers a guide for living with greater connection, fulfillment, and care, especially without defaulting to unsustainable or oppressive systems of behavior rooted in harm. Being in the flow of ongoing becoming allows us to live without the pressure of fixed outcomes. Learning to live, like language itself, is adaptive. It's shaped by shifting meanings and how we move through them.

In this way, we're navigating multiple worlds, existing in the present while always becoming ecological, toward an incomplete future state deeply connected to Earth's systems. It's an ongoing journey of growth and adaptation, even if it brings with it contradictions and uncertainties. Life's adventure is the exciting part, never a dull day when changes are happening. All we have to do is stay attuned and listen.

REFLECTING

- What practices or experiences have helped you embrace changing perspectives instead of holding onto a fixed sense of *self* about who you *are* or *should be*? How might language play a role in supporting this change?
- How does thinking about what *we* as an interconnected planet are becoming, rather than just what *you* individually are becoming, change your daily approach to living?
- How might the echo or ripple (reverb) of your communication and actions affect the people and environments around you over time?

CHAPTER THREE

Listening

Sympathetic vibration

In many social spaces, there have been increasing calls for people to stop speaking and start listening. Often called deep, active, or empathetic listening, this invitation appeals for many reasons. For starters, there's enough noise out in the world already, with political talking heads on every news and social media channel filling up space with yelling and fearmongering. There are also many opinions often rooted in divisive notions of who's right and wrong. Amid all this noise, there remains an immense beauty and need for *listening*.

As an active practice of making space for other voices, listening activates experiences, sensations, perspectives, and beliefs that provide deep learning and understanding. Although we'll stick with listening to acoustic sound, listening can also involve the whole body, as some wave forms are far too slow or fast for our ears to process. As much as it might appear otherwise, listening is a vital component of the language of ecology and the practice of behavioral transformation.

This chapter talks about language environments and communication with a bit of a sensory twist, shifting away from print and visual displays of information to the act of listening as a way of empathizing and understanding. It invites deeply resonating through *sympathetic vibration*.

The word root of *listen* branches from Old English *hlysman* to Proto-Germanic and all the way back to Sanskrit, meaning to pay attention, to hear, and sometimes to obey or to be called. It's thousands of years old. Not surprisingly, it's also embedded in the First Peoples Principles of Learning in Canada, where listening calls upon us to enter into reciprocal relations with everything around us, including "the land, the spirits, and the ancestors."[1]

Listening to the Earth; talking with animals, trees, and stones; and communicating with air, water, and even weather, as the physicist and Blackfoot Leroy Little Bear points out in his invitation to "listen to stones" speak,[2] broadens the scope of understanding and awareness and opens different channels of coexistence and sympathetic knowledge translation. So, too, in Japan, the thirteenth-century Japanese Zen teacher Dōgen spoke of observing the "voices of the river valleys," which inspired poets and teachers in his great teaching "Mountains and Waters Sutra."[3]

At first, listening might not seem relevant in a world where written language and visual media dominate public knowledge and information about the environment. But listening can be a profound agent of change in our personal lives as well as our public and ecological relations. Listening is the often-overlooked counterpart in the dance of language and dialogue. It enables understanding by forging connection and making meaningful exchange possible.

We might just consider how plants *listen* to their environment by responding to cues like a bee's buzzing, while bees interpret floral signals such as colors and scents. This reciprocal interaction ensures the survival of both, illustrating how listening and communication are essential for ecological harmony.

In the spaces shaped by sound, we can tune into the essence of things, sensing both what's present and what has faded away. Considering such ephemeral approaches to environmental communication and knowledge expands our multisensorial ways of becoming ecological, transforming our relationship to other species.

Listening to the Earth; talking with animals, trees, and stones; and communicating with air, water, and weather are a lot healthier than they might seem. There's a growing way of thinking, often called *new materialism,* that tries to reconnect the physical world with our social and political lives. It inquires into the active role matter and material forces have in shaping reality, particularly emphasizing the entanglement of the human and more-than-human world. Thinkers like Karen Barad suggest that even physical forces (trees, stones, air, etc.) shape how we think, feel, and even communicate.[4]

In this view, the universe doesn't just exist around us. It interacts with us, and we can respond with care and attention, almost as if it's speaking to us. If this is the case, as this chapter explores, then the real question becomes: *how can we learn to listen*?

Listening helps to bring the world to life. It does the opposite of what visual communications do, which is to freeze something in time, to make it into an object.

In the name of progress, we witness the destruction of entire valleys, mountains, plains, deserts, bays, islands, oceans, and every creature and plant in or on them, even when what we take winds up in a landfill shortly after. Many people collectively choose not to listen to what the Earth is saying because we often prioritize our own version of intelligence and think of the future in a disconnected, individualist way.

Based on this version of *vision-dominated intelligence*, we treat life as either having rights that protect it from intentional harm or free for the taking. With deep listening, we envision what's inside the other, what animates it, and that's akin to *sympathetic vibration*. In other words, we *feel* a sense of relation to it. This is possible even with beings vastly different from us, such as empathizing with a tree or a rock as we would with another person. Some vibrations move too slowly or too quickly for us to notice, but they're still there.

Listening attentively is often challenged by the constant presence of mechanical and electrical noise. As human environments grow increasingly louder, other species are no longer able to communicate. One clear example is how human-created soundscapes contribute to biodiversity loss on both land and in water, as noise disrupts the sensory abilities animals rely on for finding food, mating, evading predators, and so forth.

An interesting anecdote relating to noise occurred during the COVID-19 pandemic. Researchers studying white-crowned sparrow populations near the Golden Gate Bridge in San Francisco saw their populations boom. With the motor traffic at a relative halt, these small birds greatly increased the diversity of their song and were able to attract mates with songs that were otherwise drowned out by the everyday din of cars and trucks. They made a remarkable recovery until pandemic restrictions were lifted and then plunged into a deafening noisescape once again. People living in urban environments, which is the majority of the population around the planet, are constantly embedded within unrelenting soundscapes that are getting louder.[5]

The message we're trying to convey in this chapter is that listening is the quickest way to relate to others in a situated and personal way. It's about both the sounds and the silences that convey meaning and situate us as beings becoming in place. This chapter also looks at communication as a way of understanding the meaning and consequences of human generated soundscapes.

Humans have an innate ability to practice selective listening. So listening is deeply embedded not only in how we relate to each other and our environment but also how we enact ideologies and construct worldviews. To arrive at this understanding, to make it our fundamental priority, we will welcome our ears as both teachers and learners that help us to adopt approaches to becoming ecological. This chapter also expands our sense of how humans can begin, at a deeply systemic level, to advance more interconnected ways of living on this planet.

PRACTICING

How much of your environmental knowledge and awareness comes from listening? Reflect on your role in those conversations – were you contributing, observing, or both? Who or what were you engaging with in this dialogue? Do you value information gained through listening differently than what you read in articles or watch in videos? Do you find yourself more confident in facts you read or in ideas shared through conversation? By yourself or with others, choose an environmental topic that's *a matter of concern*, whether it's something small like gardening or something larger, like shifts in food production. How often do conversations about it rely on facts, and how often do they lean on personal stories or lived experiences?

Seeing is believing

The radical separation between humans and other beings on this planet reflects a troubling form of *human exceptionalism* – one that also manifests within human societies, where certain knowledges, practices, and even people are privileged over others. By setting up conditions for becoming ecological in the ways language affects actions, opportunities open to mend our relationships with all the entities that share our planet, as well as with ourselves, to halt the mass extinctions and prevent humans from sharing the same fate alongside the many endangered species, from insects and songbirds to whales and rhinos. One method to achieve this is through deep listening.

You could even say that it's through listening that we first become engaged as living persons. The word *person* (*per/son*) is rooted in Latin words meaning "through"/"sound." If we attribute other entities on this planet with the ability to make and experience sound, then it follows that we must attribute to them an inherent personhood and all the rights that human society (even if inadequately) bestows upon persons – the right to live free of persecution and harm. But to do so, we must learn to listen to all our relations and hear what they're telling and teaching us about the world.

Dominant society, through Westernized practices of consumerism and individualism, is almost exclusive in its bias toward visual culture. Just so we don't forget, we're calling this *ocularcentrism*: the privileging of sight or vision as the dominant sense navigating meaning in the world. During this period of visual bias, we have gone from expressions like "taking someone on their word" to "seeing is believing."

This domination of visual sensibility isn't as ancient as it may seem. The visual ascendency in Western cultural practices emerged alongside the European Enlightenment, a period from approximately 1685 to 1815 but that continues to influence life today. During this period, often called the modern worldview or "modernity," science and religion began to emerge as distinct domains of thought. Science increasingly focused on empirical observation and experimentation, laying the groundwork for what became the scientific method. Through systematic engagement with the natural world, knowledge was recorded, tested, and eventually distilled into what we call "facts." In contrast, the church grounded in faith, upheld belief as the foundation of truth.

But what exactly are facts? At their core, facts are distilled patterns – observations that hold steady long enough to become reliable markers of meaning. Still, when do experimental observations, bounded by control and context, cross the threshold into resolute truths? And how do such facts retain their status in the face of rapidly changing ecological conditions? In a time of instability, how do we reconcile our need for certainty with the mutable nature of the world they attempt to explain?

One of many contradictions of the modern worldview constructed mainly from factual knowledge is that it lacks the flexibility of becoming and adapting to change. Facts are the basis for concerns, but they can also function to work against those concerns. Returning to the refrain in the opening chapter, *matters of fact* have been used to argue against positive change in social and environmental *matters of concern*. This is partly why science now finds itself in a tricky bind, as the Belgium philosopher of science Isabelle Stengers has written about, caught between its celebrated role as a producer of certainty and facts, and its deeper, often-overlooked potential as a practice rooted in curiosity and discovery that opens possibility.[6]

In the book *Enlivenment: Toward a Poetics for the Anthropocene*, the German cultural scientist Andreas Weber describes Enlightenment-style rational thinking as an ideology that focuses on "dead matter," a worldview that sidelines life and living. Weber creatively proposes that we might consider *enlivenment* – both as a response and alternative term to *enlightenment*.[7] Another way of understanding an object or facts is dead matter or the unliving. Anything considered dead – fixed, certain, and controllable because it is final – is treated fundamentally differently from what's alive, which is dynamic, continually evolving.

During the Enlightenment and thereafter, discourse shifted from knowledge rooted in experience and situated listening to knowledge grounded in observation and so-called objectivity. The result was a distancing of the relationship between people and other living beings, or subjects and objects, through factual knowledge.

This transition, coupled with rapid technoscientific advancements and a significant rise in vision-based print literacy rates across global populations during the nineteenth and twentieth centuries, fueled a dramatic surge in industrial and technological development, a massive production of *dead matter*. In such a rapid time of change, the seemingly stable world of facts gives a sense certainty and control to people, explaining what they can neither see nor control.

"So what's the big deal?" some might ask. The problem with factual knowledge isn't facts themselves. It's Certainty, with a capital C. When facts are treated as fixed and final, even when they are not and can be manipulated, they tend to shut down the potential for dialogue rather than invite it. Ironically, a culture obsessed with factual supremacy can stifle the very curiosity of discovery that science depends on. Facts, presented as absolute, often feel disconnected from the messy, evolving relationships through which we come to understand the world.

When we're deeply committed to fact-driven arguments, anchoring ourselves in certainty and "proof," we often stop listening – not just to others but also to the quiet uncertainties or questions within ourselves. In a society that privileges the visual – where *seeing is believing* – facts tend to take on the weight of unshakable truth, captured in images, graphs, and headlines. This visual dominance reinforces the idea that knowledge must be seen to be real, rather than felt, heard, or co-created. It narrows the ways we engage with the world, reducing listening to a one-sided interpretive act.

In the context of something like climate action, this becomes especially tough: flooding people with facts and visual evidence often leads to deeper entrenchment, not openness to dialogue. Without listening, without space for dialogue and building relationships, the line between belief and fact grows extremely porous, and the potential for change slips further out of reach.

As already mentioned, the root of the word *listen* is to pay attention, to attend. The verb root is "to tend," as one tends a garden. To tend is to enter into relational dialogue governed by care. What's more, tending takes time. If we think about it, paying attention requires sustained listening because the focus of the "eye" is always shifting, even when concentrating, and listening also shifts the focus off the "I."

Industrial development brought about other changes as well, further challenging our ability to attend or listen, including massive increase in human populations. The presence of humanity, whether in person or as remnants in their litter, waste, and noise pollution, is very difficult to escape.

Cities began to dedicate a predominant amount of collective space to transportation, booming zones in which nothing living can linger long. In the twentieth century, cities and suburbs were designed for cars propelled by cheap and available fossil fuels, not for people to gather and talk or to forge meaning through language and dialogue. People moved increasingly inside, into sealed and hermetically controlled environments that isolated and segregated them as individuals, both visually and acoustically, and this changed the way we use language and listen.

Communications also adapted so that a physical person didn't need to move messages between these different interior spaces, it could be done electronically. In a virtual shared world, communication is action, and we move away from attuning our listening with and to the Earth. We stop listening to the rain or to the wind and creating expressions that attribute meaning to these experiences. Instead, we see only facts and beliefs as opposed to listening to possibility, attending to interconnected life.

Sounding our environment

Vision, at least for sighted humans, involves distancing objects and the perceiving subject. Humans have two eyes, and these eyes are meant to focus on a single point at various distances, a kind of rudimentary triangulation that gives information about the distance between the viewer and the focal point of what's being looked at. Fundamentally, vision separates the viewer from the viewed, and the subject from the object under observation. This division affects both our language and our thinking about the world around us.

Distance plays a central role in vision, which is inherently forward focused and linear, creating a sense of separation that reinforces the perception of objects as distinct realities. From these physiological aspects of vision, the concept of "objective fact" has emerged. In the modern world, facts serve to articulate truths derived from visual observation and the study of objects, and this is treated as different from knowledge acquired through listening and subjective experience.

Sound, as it enters the ear, stimulates the ear drum, which passes vibration to the tiny snail shell–shaped cochlea, filled with fluid. As these vibrations create waves in the cochlea, very fine hairs pick up the vibrations and convert these to electrochemical signals, which the brain perceives as sound. Sound is, however, much more plentiful than just what's perceived by the ear.

Sound waves are *proprioceptive*, meaning they're experienced by the entire body and its relation to the external world, and lower frequency sounds pass through the body and resonate in cavities such as the skull and lungs. Higher frequency waves bounce off our body surfaces, sometimes painfully, like when subjected to a screeching high sound.

Unlike vision, sound is omnidirectional and orients the listener in space, telling us where the sound is coming from in any direction. They tell us whether the sound source is large, small, stationary, or moving.

> Sound is how we understand our location in
> an environment. But just as importantly,
> sound is internalized, shaping our emotions
> and personal perceptions.

The difference between seeing and listening is simply that when we see, light waves are reflected off surfaces of objects around us. The insides of objects we see remain invisible unless the object is transparent. Even looking through pure water distorts vision. To visually know what's inside an object, we must destroy the outer surface or create a hole to peer through. By listening, we understand interiority and the essential properties that something is composed of.

One example of this is ultrasound technologies that help us see a fetus's development inside the womb without harming the mother or baby. Another example relates to mushrooms. We know that large, underground mycelial networks (of which mushrooms are the visible fruit) connect the roots of trees. But it was not until tiny electrical pulses were recorded traveling through these fibers that we understood that the trees were in communication, listening to one another through these mycelial networks.

Subtly, we interchange *feeling* our environment with *sounding* our environment when we seek to understand the inner substance of something. The case provided in Walter J. Ong's insightful book *Orality and Literacy* was of an object, like a sealed flask made of a solid, non-transparent material.[8] Unless you destroy the object by breaking it, you can only know it by its shape, its object-ness. If you sound the object, by tapping it or shaking it, then you can hear if it's hollow, containing something, or solid. You might also learn whether what's inside is solid, composed of many pieces, or whether it's thin or thick fluid by the sound it makes.

And, for another example, if you found a dark cave and could see only a few meters in, you might shout and listen for echoes and reverberations to gauge its size, echolocating like a bat, whale, or any number of species that use sound to orient themselves to what is going on around them. There are many necessary uses of deep listening to the reflections of sound that aid survival.

The focal point of human sight tends to be singular. Vision moves from point to point in the visual field connecting these points to form a "bigger picture." Eye-tracking software, for example, follows this point of focus across a surface as the viewer makes sense of an image, connecting the dots so to speak. Because our eyes are in the front of our heads, we see what's ahead of us. This informs our normative orientation and the orientation of our language of concepts and values.

Let's apply this concept to a socioecological context instead of only a biophysical one.

Western civilization is deeply committed to the concept of *progress*, which could be perceived simply as forward motion toward a goal. But progress, as an ideology and modern worldview linked to Western notions of "the good life," contains a deeper belief that human civilization can and should always move forward rapidly to achieve greater levels of prosperity at any cost. It involves the idea of continuous improvement and advancement in various aspects of life, including technology, economy, society, and so forth.

We imagine that the future is also in front of us, that time is also linear, just like our line of sight, rather than circular and all-encompassing, as it would be if the concept was modeled on listening. As the drive for progress oftentimes overlooks or undervalues sustainable practices, it can also lead to environmental degradation, social inequalities, and cultural assimilations because we don't tend to look behind us at civilization's wake.

Progress and linear time, always moving forward, puts the past behind us. We no longer see it. And unless pressured to do so, ideologically we tend to favor forward thinking and forward motion, being at the cutting edge, at the forefront, being the first in line. This is definitive of much of contemporary society. Advances in technology, medicine, industry, and so on are about moving forward from our current state. This urge to get ahead and to put the past behind us has had dire consequences. In our race to be the leader of progress, many ethical and moral questions are continually being put aside or even ignored completely.

*Now, what if we shifted the focus to sound and listening
as the dominant sense and communicative medium?*

Vision is projective;
sound is immersive.

Now, just for a moment, consider how our understanding of the world might change if society prioritized listening over seeing. In this alternative reality, perception would become more fluid, attuned to nuance, and rooted in collective awareness rather than fixed visual snapshots. Where vision is unidirectional, requiring one to turn one's head to see objects that are elsewhere, sound is omnidirectional. Having ears on the side of our head means we hear as well from behind as from in front.

Sounds of objects aren't separated from sounds of subjects. Instead, they exist as a symphony of interconnected, resonating beings. Sounds aren't frozen in time either, like stills in a movie of life. Sound conveys the temporal continuum. When a voice stops speaking, it recedes into memory. Whereas when light is captured, it can be held still as a representation, as an object like a photograph or a visualization of data. Sound is alive.

This idea that we can freeze time and space governs the notion of objective fact, particularly that something can remain true regardless of how the context and environment changes. As an internalized sense, sound is always embodied communication with the world, meaning it transcends spoken language. It gives us a subjective positioning, a world that's conjured by resonations with other subjects. It's intuitive, sensing and adapting to external stimuli like an ecosystem.

When we say something *rings true*, for example, we mean that something resonates with us, it makes sense intuitively, even if we don't have objective facts to back it up. And this ringing of truth is a *felt-sense*, a resonation that originates not solely in the brain, but throughout the physical subject.[9]

We've been focusing on vision as a contrast to listening to show how listening provides a more connected and ecological way of understanding and how it subtly affects the ways we communicate. When imagining a world dominated by listening, we stop seeing and cataloging all our differences. Instead, we hear our commonalities and feel through *sympathetic vibration* what makes us similar to the other subjects in the world around us.

Language in this context brings us together. In this view, the world is alive with personality and connection, and our language would naturally reflect this connective way of being. We would start to recognize the personhood of all the so-called objects in the environment and the possibility of becoming in dialogue with them.

EXPERIMENTING WITH POSSIBILITIES

Imagine a world where hearing, not sight, is our most important sense. In this society, sound and resonance shape everything. This is how people communicate, learn, and navigate their surroundings. Buildings are designed for the best acoustics rather than visual beauty, with rounded domes enhancing sound. Stories and traditions are passed down orally rather than through writing, and progress is measured by how well communities and ecosystems stay in harmony and balance rather than by material success. In a world centered on listening, would people value connection and shared understanding more than competition? How might our relationship with the environment change if we focused more on sound than sight?

Living in sound

This brings us to the concepts of resonating and resounding, or as we introduced in the Becoming chapter, *reverbing* – a way of becoming ecological through *sympathetic vibration* that can activate multiple senses as an embodied language. If we were to simplify this, we could say *connecting with nature through vibration*, but this does leave out some important details of the experience.

Just to recap, reverb is the phenomenon of sound waves bouncing off surfaces, adding a fullness of tonal textures, like light through the facets of a cut gemstone. Reverbing is how live sound is experienced. For example, many people dislike how their voice sounds when they hear it recorded and played back through speakers. It often sounds different from what they hear while speaking. And they're correct because it is different. No longer is the voice resonating internally, in the skull and chest cavity, where it takes on the characteristics of our personhood.

In the language of ecology, the concept of *deep* has distinctive connotations. While sometimes it means *far into*, like *deep woods* means far into the forest, it's often used as an adjective to describe approaching a topic in an embedded and embodied way through multiple senses. It means committing to actions that go beyond superficial gestures in our conversations, as the term *deep ecology*, coined by Norwegian philosopher and former Greenpeace Norway president Arne Næss, suggests.

Deep listening holds the same kind of implication: to listen with full presence, paying attention and holding space for sound without labels or judgments. It allows for greater possibility. But what does this mean for the language of ecology, you may be wondering? Well, beyond the need to listen more deeply to the world around us for signs of its health and to care for it, listening can also improve our conversations and understanding through dialogue.

Listening deeply helps us to envision, which is different from seeing and closely related to insight. Listening more readily conveys understanding, the path of wisdom, which is different than knowledge. The turning toward Indigenous and ancestral knowledges in ecological and sustainability work, such as in Traditional Ecological Knowledge, represents an attempt to balance runaway object-knowledge systems so practiced in the Western world with the slower energies of understanding and wisdom rooted in traditional practices.

Some Indigenous traditions use ceremonial healing circles, also sometimes called circle work in other non-ceremonial community, health, and educational contexts, where people come together to listen and receive members of the community. A symbolic item, such as a feather or stick, is passed from person to person. It represents the turn-taking sequence of speakers so that the majority of the focus is upon the act of listening deeply, not upon the performance of speech. The speaker isn't concerned with making an argument through speech. The speaker is witnessed, as they express their subjective truth or experience at that moment, which is different from arguing over what are perceived as facts. It's an act of holding space for multiple perspectives, a diverse ecosystem of ideas, experiences, understandings, and so on. Listening, in this context, is the leading dance partner in the exchange of listener-speaker.

Listening is how we slow down and actually tune in, like switching off autopilot and really paying attention to where we are and who (or what) we're with. Listening is the opening to this slower, situated, and responsible practice of living. It's the way of *envisioning* – seeing with more than just our "eyes" or "I."

Resounding empathy

When, socially, people say they aren't *being heard*, they're emphasizing a lack of empathy and relating (co-creating meaning) rather than a lack of visibility. This is partly why talking circle work exists. It's possible to recognize someone as an individual but still not treat them as a person and a living being.

To be a person is to be heard, as the word *per* (through)/*son* (sound) implies. If we remain silent to the world around us, we become lonely on a crowded planet – surrounded by endless knowledge and information but lacking the understanding to change the course of dominant Western progress. Listening to the diverse experiences of people who lead very different lives from our own can change our behavior patterns and our ability to empathize on a personal level. Personal understanding often sparks activism and drives us to make choices that can initiate meaningful change.

Listening is a key way to *affect* us, activating how we feel and behave in relation to other persons (humans and more-than-humans). Anyone who has ever done the experiment where you take a horror movie and at a peak fright moment, turn the sound off knows this. Suddenly the terrifying becomes just another scene of something unpleasant or even comical and absurd. It's the soundscape that grips us in fear, that foreshadows and unnerves us by suggesting what's outside the visual frame. The sound makes it seem real, conveying the felt-sense of horror. The same is also true for tear-jerking dramas or even environmental documentaries. The sound activates our tears and empathy while the drama unfolds on screen.

In environmental news cycles, people often encounter distressing images that can lead to compassion fatigue. Think of any number of fires, floods, hurricanes, and typhons around the world. Just over the past ten years, they're too frequent to name and remember at this point.

The Canadian Medical Association defines *compassion fatigue* as "the emotional toll of caring for others and witnessing their suffering while wanting to help alleviate it."[10] This phenomenon is widely studied for its impact on behavior, with conversations ranging from the discouragement caused by overwhelming disaster news to the potential apathy fostered by presenting facts in a comforting or sugar-coated manner.

But competition for *attention* among environmental causes and issues creates the need not only to compete among themselves but also to compete with the whole marketplace of information and entertainment. As the public switches attention from one big issue to another, more sensationalism comes into play, and compassion burnout increases because of this competition for attention. Each person will have a limit to the number of issues they can relate to emotionally and dedicate their care and support to.

This is relevant to how we listen. For when we listen deeply, we are drawn toward deeper ways of relating. But the excess of visual pleas for attention produces a sensation of overload and subsequent *desensitization*, producing a numbing effect to concerns we should be able to feel.

There's no way to give listening time to everything needing attention. The constant exposure to traumatic imagery and alarming messages, such as those depicting ecological crises like extinction, devastation, loss, flooding, and habitat destruction, can saturate and overwhelm people. Without the emotional resonance of spoken words, the *affective* feedback of hearing matters spoken, this saturation often leads to feelings of hopelessness and powerlessness to enact change.

Listening, by comparison, increases personal connection to issues through sympathetic vibration and empathy generated through dialogue. In Greek, *pathos* is a word meaning appealing to emotions, particularly misfortune and difficult-to-express feelings, to relate to another's experiences as if they were your own (as in the word *sympathy*) or as being important for them (as in *empathy*). Compassion fatigue is a kind of visual overload, literally knowledge without understanding and feeling without empathy, where the agency of care is buried under the assumed inevitability of objective facts.

The universe is speaking to us

It's true. The universe is speaking to us right now, if we only had time to listen. The world of listeners seems impossibly slow when one has been immersed in computing clock times and vision dominates reasoning their whole lives. The speeding up of our sense of time brings with it impatience, the desire for the immediate fulfillment of our goals or desires.

Deep listening is impossible if we're feeling impatient and rushing to catch up. Haste makes us want to move faster in a particular direction. Deep listening is best when we're still. And when we're impatient, the whole world becomes object-oriented, and vision – the closest relative sense to rapid time – takes over. We don't have time to listen to each other or to the world. Visually, we can witness great speed with pleasure.

Vision loves spectacle. We can watch explosions that would terrify us and find them aesthetically appealing, like the mushroom cloud of an atomic bomb explosion. We can play games that entail murder and destruction and feel a thrill through this kind of visual entertainment, drawn in by the upbeat soundtrack and fake-sounding grunts and death groans of characters.

Western societies have normalized this kind of ocular-centric worldview. Notice even the word is world*view*, not world*sound*. The lack of relationality that results is no longer even a matter of concern. Language and meaning have altered in ways that deeply affect our relationships and ways of becoming. When increasingly disconnected and anxious about the future, we don't seem to know why. Perhaps we should pause and just listen deeply.

One way to interpret this ecological anxiety is as a result of relying too much on vision to understand a constantly changing world. Perhaps the kinds of dialogue we are promoting in this book – our conversations with living beings – are happening all around us right now, even though we don't hear them.

Is it possible that the world is resonating with us right now, everywhere? Is this what we mean when we say we're already ecological?

Even solar masses, like planets and suns, have person-hood. They're not inert, even when we find no signs of biological life. If being a person is being in correspondence with others through sound, then it's time to reframe who and what is called a person, even if that person inhabits a radically different "clock" than our own. It may seem surprising, but all stellar bodies emit paradoxical inaudible sounds, some of which can be listened to on NASA's sonification website.

Sonification is a relatively new scientific method that turns otherwise inaudible data into sound and employs listening to understand data. It differs from other kinds of scientific practices because it makes data more visceral, felt through our body's senses. It returns subjectivity to the objectified world. By converting long, slow waveforms into audible sound waves, we can understand phenomena such as tidal motion, glacial melt rates, or the personhood of stars in a completely different way. Sonification is also now contributing to climate science, returning to scientific data the affective, temporal, and relational qualities they have been missing since the Enlightenment, when we began to favor vision and stopped listening to the Earth.

The Earth also emits a sound, a frequency of 7.83 hertz (cycles per second), which is too slow for human hearing but connects us all, everything and everyone. It's always present. Recognizing this means understanding that the planet, as an integral whole composed of many interconnected systems, is alive, and in the grand scheme, is also a per/son. If we slow down enough, we might start to hear it and be in dialogue with the whole Earth system. That, in a nutshell, is what it means to become ecological.

Even if we can't *hear* the language of slow frequencies, we must learn to *listen* to the Earth to heal it and ourselves. In this way, we open ourselves to a deep and life-affirming kinship mediated through the language of sound and the practice of deep and slow listening. Such a language requires new forms of learning, more expansive approaches to ecology than are taught in science classes, and layered forms of communication that co-create meaning among all relations, whether human or not.

In doing so, we cease to view the world as an object, something to be taken or exploited for temporary profit and ongoing convenience. We might grow to appreciate and respect planet Earth as the best company any being can have. The Earth is alive and always listening.

In this calm state, we might no longer feel fatigue from our exhausted compassion, inundated by visual bombardment. But, instead, we could start to feel invigorated and live collectively with a sense of purpose that connects us to the environment and the slow time of evolution and sees those relations as the source of life's purpose and meaning.

Although there are no quick fixes to the many *matters of concern* in the world today, listening, or the lack thereof, is implicated in all of them. Deep and collective listening could ignite a lasting, generative healing of the planetary spirit. Listening also is fundamental to how we relate to everyone and everything.

REFLECTING

- How can deeper forms of listening promote greater care for the planet? What about for one another?
- How might shifting from a visually dominant way of understanding the world to one centered on listening change the way we perceive and interact with our environment, other beings, and the concept of progress? Can you think of an example in your life or community when *listening* rather than *seeing* could foster greater connection or understanding?
- Can you think of an instance when deep listening changed your perspective, worldview, or motivated you to act in some way?

Relating

Connecting the dots

Imagine walking into a crowded café where no one is speaking, not because it's silent time or people are deeply listening, but because everyone is hooked to their screens. Some could be messaging friends, while others might be reading news or chatting with someone across the world. It's a snapshot of how our sense of relationships today stretch across space in new ways, blending physical and digital presence. And yet it raises an age-old question: how are these evolving forms of connection shaping who we are becoming in every moment, and what do we understand as being together in this collective emergence?

Whether digital or in person, the truth is that humans are very social organisms. An important aspect of this is how the quality of our lives directly reflects how we value and engage in relationships. Our interaction depends upon each other not only for support, care, love, and so forth but also for the ways we construct our realities about who we are and how our identity and thoughts are formed through language. These ways of interaction have a capacity for social interaction over individualistic meaning. The latter, however, has dominated society in how we talk, think, and act.

In the Western world, we often approach relationships and the languages that support them as by-products of individual identity. For instance, there are over 2,000 words for individual expressions in English, with a fraction of those for relationships. Language use even affects our thinking, including our perceptions of and responses to reality. Whether or not it seems obvious, all thinking emerges out of our social environments and our social environments emerge through language.

Lev Vygotsky, a foundational Russian cultural psychologist and linguist, claimed that all sites of meaning come through relationships.[1] Others have similarly believed that the answer to the grand question of life's meaning lies in the value of relationships. So the relational fabric of our lives is no trivial matter, whether it's made of a virtual, imaginative, or physical material.

But many ways of communicating and making meaning divide and separate into individual units more than allowing interconnection and showing interdependence. These language systems create feedback loops that shape realities and limit our ability to see ourselves as beings with multiple identities emerging from our social contexts, rather than as isolated individuals. Following this premise even further, the language that creates meaning shapes how we perceive ourselves and the world, influencing our ability to see or even listen to everything as interconnected.

How, then, might we enrich practices that mend the rift between "I" and "we" or "self" and "collective," acknowledging how this unnecessary separation fuels the fragmentation of humans from the living systems that sustain us?

This chapter looks at how the ongoing process of becoming ecological hinges on relating with other humans and all our relations (more-than-humans), including animals and other organisms and entities that share the planet with human species. In the words of Anishinaabe Melanie Goodchild, relating is about our relationships with the lands and everything living that connects with it.[2] In other words, relating is another type of ecology that is rooted quite literally in our environments.

So what does this really mean? The ordinary language we use, embedded in larger social discourses, has significant influence over our perceptions and behaviors. Put another way, language affects the process of relating. Calling a person a "resource" instead of a "person" will undoubtedly shift how we view them. Likewise, labeling a 300-year-old ponderosa pine tree a "resource" alters our sense of connection and value to it.

Meaning all depends upon how language is framed and perceived, as well as who controls the overarching narrative constructed out of language. The encouraging part of this is that many cultural paradigms already exist that highlight relational forms of language and behavior, many of which also happen to mirror ecological systems.

Understanding how things are connected is crucial when dealing with ecologies, where systems are basically made up of the relationships between their parts. Recognizing these connections helps us grasp the idea of *interdependence*, rooted in the Latin words for "among or between," where relationships form the essence of the system.

Take the experience of *connectedness*, for example. It involves sharing information across systems. By acknowledging how everything is interconnected, we broaden our focus beyond individual perspectives and enhance our emotional and mental engagement.

To capture the scale of interconnectedness, we might think of it like this: we're part of smaller groups like communities and larger entities like the Earth. It's another paradox – like being a piece of a puzzle that fits into both small and large pictures. Imagine that the Earth is an electron, rotating around a proton (sun) in a molecule (solar system) that's clustered together as a cell (galaxy) in the body of a god (universe). Then imagine that in your body there's a cell, and within that cell made up of molecules there's a proton with an electron that's covered in life-forms just like Earth. Suddenly, the "I" quickly dissolves into a larger constellation of "we" that includes everything.

———— ∘∘◯∘∘ ————

This nested relational model goes on infinitely in both directions. The macrocosmic and microcosmic aren't separate but constitute a continuous system of interdependent relations. The astronomical metaphor referred to above helps us imagine how deliberate shifts in language use at the micro-community level can support change at the macro-social level.

When we learn within these nested systems, we can see that relationships shape our understanding of the world as part of many worlds, many systems with many dimensions of interaction between them. Models of traditional education often stress the importance of individual thinking, where one person called an "expert" knows something and shares it with the hope that others can receive and embed it into their own thinking.

But what if we shift that to a more collaborative approach, where we explore knowledge together? This change, known as *dialogical thinking* because it's in exchange with each other, helps build better relationships with, in, and among social and ecological systems. It's not just a by-product of working together but also the very essence of it.

Connecting with others, using language, navigating experiences, and relating to a more-than-human world – all of these interactions shape and sustain ecological systems.

Humans and everything else, like animals, insects, stones, and plants, are crucial parts of these systems. Together, through our interactions and experiences, we create meaning within these systems. Meaning arises from this shared responsibility.

But before delving into the topic of *relating* as a language ecology, it's useful to look at individualism as a construct of Western societies, especially the contested concept of the "self," which contrasts with other perspectives emphasizing interconnectedness with others. We then explore relational paradigms such as responsibility and kinship.

Beyond the self

We don't often link conversations about the "self" to socio-ecological change, apart from the usual focus on individual actions like recycling or purchasing an electric vehicle. But looking at the self is a deeper dive into language and identity, and it may be one of the most significant barriers to embracing an ecological mindset.

> *Self* is all about you;
> *ecology* is about everyone and everything.

Most would likely agree that the dominant paradigm prioritizes the individual self. Look at the billion-dollar "self-help" industry or the countless TikTok and Instagram posts urging people to discover your *true selves*.

Who are we, really? Who is this "true" or "higher" self people talk about, and why do we never seem to find it? Or if we do find it for a fleeting moment, why does it keep changing depending on the relationships around us? Is there an ecological self? Why do people cling to a single identity when we are made of many experiences connected to a bigger network of ecologies? There's a path that led us here, and it holds many clues that can guide us away from it.

The Western paradigm of individualism, rooted largely in Greek, Roman, and Christian traditions, gained significant momentum during the Enlightenment period from the late seventeenth century to the early nineteenth century, prioritizing logic and reason over organic or cosmological worldviews. This cultural and intellectual shift aimed to challenge corrupt systems enabled by monarchies and the church, with the aim of promoting freedom through democracy that enhances cultural diversity. Previously, for those connected to the European continent, science and rational thought were often suppressed, but their rise led to advancements in things like technology and medicine.[3]

The French mathematician and philosopher René Descartes's famous statement, "I think, therefore I exist,"[4] not only epitomized this era but also influenced the future we live in now. This seemingly simple concept reinforced the division between mind and body, planting the idea of a self that stands apart from the living world it depends on. It also encouraged *reductive thinking*, the practice of dividing knowledge into separate parts to grasp reality. Other influential Western thinkers like Isaac Newton and Francis

Bacon extended this approach, which often overlooks the living interconnections that bind systems together.

What's now called *Cartesian dualism* (based on Descartes's Latin name) separates the mind and body, or thinking and matter, influencing modern culture and shaping society's focus on the individual. This paradigm of Western "modernity" has more broadly viewed the mind-thinking self as distinct from other beings and their surrounding environments, encouraging the perception of detachment rather than one of interconnectedness. As the anthropologist Philippe Descola puts it, "modernity set itself up in a cloud, way above ground, by claiming to separate humans from non-humans, nature from society."[5]

While the mind–body split remains entrenched in Western culture, another barrier that emerged historically, even before the Enlightenment, stems from language. As pointed out by the Canadian media visionary Marshall McLuhan in his somewhat forgotten 1967 cult classic *The Medium Is the Message*, 3,000 years of Western history has been largely shaped by the introduction of the phonetic alphabet.[6] This shift from oral and auditory communication to visual language depends largely on the eye (through vision) as a way of breaking reality into discrete units and constructing architectures of knowledge with those bits.

Alphabets, for instance, require a linear construction by putting fragmented bits of sound together to visually reconstruct speech. The script freezes speech in time, makes it possible to separate speech and speaker, just like the mind–body separation of Cartesian reductionism. Alphabets shape our perception of environments by favoring visual representations of time and space as uniform and continuous, akin to the linear construction of sentences or paragraphs. Even as we compose text in this alphabetic medium of a book, the structure inherently follows a linear progression so that the architecture of knowledge we're sharing is revealed as internal dialogue.

And yet this book you're currently reading also subverts this linearity with chapters arranged interchangeably and employing language and action in a recombinant way, reverbing and overlapping, cycling through networks of meaning rather than singular topics. We're doing this intentionally to reflect the complexity of the environmental world represented in discourse ecology.

The medium of the English language doesn't fit as well with nonlinear structures, so there's another paradox here, where linear language must be used to communicate nonlinear concepts. We use the pronoun *we* to underscore our *polyphony* in linear text, the many voices that are speaking simultaneously like when we go outside and listen to the birds the insects the winds the leaves all speaking together.

McLuhan's notable observations about printing and the alphabet, although not commonly associated with ecological perspectives, underscore how societies' dependence on alphabet literacies and their visual orientation contributed to the widening gap between humans and their environments through language. This gap is associated with a history of discourse.

McLuhan was also, incidentally, the first person to popularize the notion of *the information environment*, meaning an environment that one virtually inhabits and yet has direct consequences on how people interact with the physical environment. This concept is closely related to what we refer to as discourse ecologies in the Discoursing chapter.

Taking this a bit further, the notion of *reductionism* – that things can be broken apart in pieces and disconnected from each other – emerges from organizational thinking of alphabet literacies. This ties to rationality and logic, which serve as Western philosophical foundations for organizing thought and language into structured sequences of meaning. The technology of the alphabet, as McLuhan notes, relies on the rationality of visual experience.

Such a practice stands in contrast to acoustic space, which is defined by expansive and omnidirectional encounters. As we explore in the chapter Listening, it's a space shaped by the surrounding or immersive experience rather than by a linear *line of sight*. When relating this to individualism and its impact on ecological awareness or existence, this cultural narrative isn't only dominated by the "eye" but also by the personal "I" of everyday experience.

Questions of the "self" have arisen to ultimately answer larger questions about reality – who we are, why we exist, and what is the nature of our reality as humans?[7] One of the shortcomings of human evolution, particularly in the modern age, is that we've created some arbitrary separations between our biological existence and the individual thinking mind. Such a separation prioritizes a self as central to building society, influencing our decision-making and even our sense of reality.

The self, also called the "I" or "ego," is often perceived to be somewhere inside of our head, the master puppeteer controlling our every action. This abstract self creates the conditions for reductionist thinking, such as when we reduce the value of something like a steelhead trout to its abstract, monetary equivalent as a resource to be bought and sold at the market.

When asked who we really are, the answer immediately comes back to our sense of self. "Why, I'm *me*, of course. I'm special and unique – one of a kind." But is our sense of "self" or "I" real or only a constructed made-up reality, a story that has been built over the years to make sense of sensations and experiences? And how has the deeply rooted belief that we even have a self ultimately contributed to the collapsing of ecological systems? It's heady stuff, no doubt.

Alternative perspectives exist that view the world holistically, recognizing the interconnectedness of all parts and the importance of considering systems as a whole. Perhaps the question might be better shifted from *who we are* to *what are we part of*? What if the self isn't a singular thing or *entity* at all? Could it be instead an act – a way of becoming through change, much like how language contains both a sense of grounding and movement?[8] A verb, not a noun. These perspectives, as seen later in the chapter through the notion of kinship, are essential to adopt for addressing many ecological challenges.

Here's the tricky part. Our understanding of ourselves is shaped by past relationships and experiences, indicating that our sense of self has been constructed in society through language rather than something that's inherent to our existence. Consider, for example, how our sense of identity is shaped by the beliefs and norms of the social systems that surround us, such as our family, friends, social media platforms like TikTok, religious communities, mentors, or educators. The concept of self is fluid and constantly influenced by the discourses present in these *information environments*, which contribute to its formation and perpetuation.

Individualistic language that focuses on a singular pronoun of "I" or "me" further entrenches the power of the self. Instead of thinking or saying, "I experience fresh air in the forest," consider it another way: "the forest provides fresh air." This subtle but powerful shift changes the identity from you to the forest, emphasizing the relational aspects of sharing the air in the environment.

EXPERIMENTING WITH POSSIBILITIES

Imagine you're a drop of water in the vast Pacific Ocean. You initially have the sense of being a distinct drop, unique and singular, but when you look closer, you realize that you're not separate from the water around you. You're constantly mixing, merging, and transforming with the waves, currents, and the other drops. The boundaries you once thought defined "you," as a distinct "self," are simply temporary and ever-changing. This little experiment in thinking challenges the notion of a permanent "self" by illustrating that identity might be fluid and interdependent, shaped by countless interactions rather than something fixed or independent. It raises the question: if there is no clear boundary between "you" and everything else, does a distinct "self" truly exist?

Reconsidering the self and its supporting language practices is closely tied to ecological principles and may even be a key impasse to forging a deeper connection the environment. The belief in an individual self often fuels consumerism, for instance, prioritizing personal desires and pleasure over sustainability and driving overconsumption. This entrenched mindset hinders ecological relationships, as people cling to their identity and privileges that may come with it instead of considering the needs of various environments around them.

Another less obvious example of the belief in a "self" that's hindering ecological balance involves how people often attach their identity to certain beliefs or ideologies. For example, a denier of climate change may reject scientific evidence due to their political identity and party affiliation, while a climate activist might overlook the need to collaborate with energy companies, despite their role in both causing and reducing climate issues through renewable energy technologies. In both cases, identity can overshadow various opportunities for connection.

While the concept of self is not typically linked with *confirmation bias*, which involves seeking out and interpreting information that aligns with existing beliefs, the two are directly connected. Without being attached to a sense of self, we're less constrained by entrenched beliefs and attitudes that affect our decisions and behaviors in relation to others. This connection largely comes from shifting perceptions – from focusing only on ourselves to understanding that all life is connected. Simply observing the intricate beauty of a rainforest or high desert in spring bloom eliminates a sense that we are separate; we're one with that experience and sensation, integrated with the unmediated aesthetic arrest of our senses.

> *How might we refer to ourselves as interdependent*
> *beings in a larger system of interconnection?*
> *If we're not a static "self," with all the individualistic*
> *baggage attached to this concept, then what or*
> *who are we to become?*

One possibility as a counter to the reliance on self is the increasingly used term *person*. Another is the term *being*, with its dynamic interplay with the process of *becoming*. Both recognize the multipart and socially embedded nature of humans and more-than-humans. We're not a detached, individual drop of water in an ocean separate from other drops of water. If we're all ecological beings that are part of the larger ecosystems of the planet, then how can we reconcile a separate sense of self in this process?

To end this section, let's invoke the prophetic words of Canadian Shakespearean actor Douglas Campbell: "When there is nothing left to burn, you have to set yourself on fire."[9] Rather than a call for destruction, this phrase can be read as a relational invitation to transformation. In the context of discourse ecology, perhaps it should be reversed: *before burning everything that is left on the planet, you have to set your "self" on fire.* Like all great fire metaphors, such as the phoenix, renewal isn't just a final outcome. It's an ongoing practice, a continuous process of becoming ecological.

Collective responsibility

Discussions about ecological issues tend to quickly focus on personal action and responsibility. "Am I doing enough?" But that question has no clear or satisfying answer. What even counts as *enough*? How can that be measured?

Perhaps a better way to approach this is to recognize that many of us feel an increasing weight of responsibility. This burden can seem overwhelming for two opposing reasons: we're taught to see ourselves as separate individuals, yet we're also deeply connected to the spaces and places that shape how we see the world. And many of us likely feel this palpable tension in our daily lives?

In the remaining sections of this chapter, we highlight that being responsible ecological beings in our communities and on the planet goes beyond just individual thoughts and actions of measuring if we're doing *enough*. When we focus too much on individualistic language, it holds us back from meeting significant challenges. And this all ties into how we shape our understanding of the world through collective language practices that affect the worlds we create and live in.

Shiela McNamee's and Kenneth Gergen's book *Relational Responsibility: Resources for Sustainable Dialogue* provides some useful guidance into how words and language extend our social interactions, shaping relationships through the ways we engage with and use language. Words such as "trees" or "plants" serve as navigators for our understanding of the world, but their effectiveness in this role relies on the cultural contexts in which they're used, whether as references, guidelines, imaginative constructs, and so on. Their meaning relies on a collaborative understanding and corresponding value.[10]

Trees might signify a place of solace or refuge, a part of a thriving ecosystem that sustains life of many living organisms, a resource for constructing buildings for human habitation, or a part of a neighborhood that shades gardens for growing food or protects places where people sleep. These meanings are derived from our social contexts, not from individual understandings, and are dependent upon how these social meanings are created and valued. The aim, then, is to create a relational language that holds both rhetorical power of persuasion and practical function.

Instead of saying "trees provide resources for me to live," we could say "we're partners with forests in sustaining life on the planet," emphasizing mutuality and interconnectedness. This suggests that people are parts of larger ecosystems, and our sense of identity is shaped by them. How different would it be if we used the term *teacher* to refer to old-growth forests?

Assigning individual responsibility separates a person from the ecosystems they're part of, making blame unfair and inaccurate. Relational responsibility offers a useful approach for speaking and acting ecologically. A good starting place might be to enhance perspectives in our social circles, schools, communities, and eventually society at large that recognize our interconnectedness. By embracing these relations, we can develop a sense of responsibility that goes beyond the individual, acknowledging our combined impact on the world.

As an example of relational responsibility, we can look at a coastal town coming together to clean up a polluted beach, with everyone contributing to the restoration of the shared ecosystem. Instead of dismissing responsibility with "not my problem," recognizing shared responsibility emphasizes the need for a collective response to pollution, regardless of its origin. As residents of the town work together, they not only remove litter but also build deeper bonds within their community and with the environments around them. In doing so, they also develop shared languages of dialogue around collaboration as a practice of shared responsibility.

By recognizing the interrelated impact of our actions on ecosystems, relational responsibility encourages collaborative efforts toward environmental stewardship. Instead of finding an individual target for blame, a scapegoat, we might instead understand how our shared connections sculpt our lives.

Phrases like "do your part," "you should take responsibility," and "you're not doing enough" all point to the individual as the primary source of change. Relational language might activate responsible action through verb structure, as we'll see in the next section, such as partaking, sharing, and contributing to shift cultural practices. This language provides a unified invitation, a sense of belonging that ultimately drives social and ecological change.

How then do we make such a huge shift? The value we place on relationships to the environment is a start. But even more so is how to perceive ourselves as relational beings in a state of perpetual becoming, connected to each other and the more-than-human persons that comprise the greater *we* as citizens of the planet.

Kinning as relational ecology

A primary reason for this chapter is to explore the idea that talking and acting ecologically can enhance collectivity and, ultimately, forge deeper relationships with our many surrounding environments. In doing so, we shift how we use discourse once again. This section explores how shifting settler-colonial legacies of human primacy over more-than-human *kin* or *persons* or *beings* remains significant to ecological language and communication.

As an entry into this language ecology, we turn to Potawatomi ecologist Robin Wall Kimmerer, who invites us to consider "learning the grammar of animacy" – or a language that animates the natural world to everything as living. Once something is living, in other words, it changes the relationship we have with it. In Kimmerer's teachings, living systems are described as using animate language, seeing them as beings rather than objects.[11] This is one reason why an economic system like capitalism uses the term "natural resources," what at one time was called "natural capital," owning up to its real function.

The term "resources" isn't truly "natural," of course. It's a concept shaped by society, and rooted in the narratives of resource capitalism, that turns language into something objectified, disconnecting it from living systems. In resource capitalism, the economic analysis often falls short of including the real costs of extracting resources and dealing with effects like pollution, displacement, and natural disasters. These costs are more than just financial. They occur because natural resources are treated as if they don't have any associated costs in economic calculations. Natural resources also reinforce national identity, symbolizing natural rights of freedom for dominant global cultures.[12]

While some may quibble about this point, it's clear that capitalism is more than just an economic system. It also functions as an ideology or a belief system influencing how we talk, think, and act. Because capitalism originated arguably in England back in the sixteenth century, with an increased expansion of it in the eighteenth century alongside Enlightenment thinking, it still serves as one of the more influential belief systems for expressing concepts in English.

According to a research study, which analyzed 1.5 million English-language books using a Google Books tool called Ngram Viewer, the English language has changed overtime to reflect capitalist values. Specifically, it showed a consistent increase in the use of words like "get," "unique," "individual," "self," and "choose" over the past 200 years (1800–2000). Over the same period, words like "give," "obliged," and "belong" decreased. The findings indicate a historical shift toward individualistic perspectives during these two centuries, underscoring the influence of specific ideologies on language use and cultural meaning.[13]

English can feel disjointed, emphasizing the focus the "self" through pronouns and nouns that prioritize individuals over collectives. As the most spoken language in the world, its global reach is deeply tied to British and US expansion.[14] Some call it imperialistic, and with good reason. English has played a major role in erasing Indigenous languages.

This history contrasts with an ecological perspective that recognizes the symbiotic relationships among all living beings. While not impossible, expressing interconnectedness in English can feel limiting, as the language often lacks the fluidity to fully capture relational ways of being.

An encouraging perspective on this is that other ecological discourses can inspire English speakers to shift away from the belief that humans are separate from their surroundings. Kimmerer openly shares this discovery in her roundly praised book *Braiding Sweetgrass: Indigenous Wisdom, Scientific Knowledge, and the Teachings of Plants* while relearning her nearly extinct ancestral language of Potawatomi as an adult. Potawatomi is a dialect of the Anishinaabe (or *Anishinaabeg*) language, which was deeply impacted historically because it was forbidden to speak in forced-English Indian residential schools.

As noted by Kimmerer, English contains around 70 percent nouns. In contrast, Potawatomi is made up of over 70 percent verbs.[15] Why is this such a big deal? Because nouns are static, unmoving words that essentially objectify a person, place, or a thing. If we just break these down, a person, place, or thing is equally living (even though some might disagree with this claim). A person or being is obviously a living entity, not an object. That has been well established at this point. But that's where many may stop.

A place, such as a biodynamic ecosystem consisting of forests, rivers, species, fauna, flora, and organisms, is also living. A "thing" gets even trickier to define simply as a noun. A thing has historically been used to refer to everything that isn't human – what many now refer to in the affirmative as *more-than-human*, implying the expansive aspects of the living world. It's a discursive move to call something a noun, and particularly a thing, because it automatically categorizes it as an object and therefore something to be objectified or othered as *lesser-than-human*.

Referring to a forest as "an untouched resource" reduces it to a noun-object, ready for exploitation, rather than recognizing it as a living, dynamic system of trees, animals, fungi, microorganisms, and rivers, all interconnected and vital to ecological diversity. In English, people express this non-personhood of things by referring to them with the pronoun "it," absent of value as an object, a thing devoid of personhood. There's a large ideological divide between the relational "we" and functional "it."

Nouns are like the separate parts of the Cartesian mechanistic system mentioned earlier. Nouns isolate pieces for dissection and definition, as well as for control and ownership. This sociolinguistic dynamic, whether or not intended, has had a deleterious effect on human relationships with the ecological world.

Kimmerer laments: "The arrogance of English is that the only way to be animate, to be worthy of respect and moral concern, is to be a human."[16] A language built on nouns constructs a world of composite entities existing in isolation. In a similar fashion, physicist David Bohm also claimed that nouns are falsely misleading because they represent unchanging isolated things that are antithetical to the ecologies of language or life.[17]

Western culture, where English remains the dominant language, might see the world in a fragmented way because of a reliance on nouns instead of verbs, just like a reliance on *seeing* instead of *listening*. In many other languages, actions (verbs) don't have to be set in action by things (nouns). In Japanese, rather than saying, "I dropped the pot," it's common to say, "the pot fell" (鉢が落ちた – hachi ga ochita), shifting focus away from individual agency and emphasizing the action itself.

In English, it's the opposite. The subject (noun) is typically the doer of the action (verb): "He broke the vase." This places full responsibility on the person rather than the action. While this may just seem like a silly academic puzzle, it reflects a deeper cultural tendency to prioritize objects over processes, nonliving over living.

This vicious cycle doesn't have to be permanent, however. Verbs can also be used instead to describe more-than-humans. This has been referred to as *personhood*, or the identifier of legal rights to living organisms, such as a giraffe, river, forest, or lake. As Kimmerer noted through her storied account of relearning the language, *a river* is now *to river* or *rivering*. Such a shift in discourse animates the *object* of a river to a *person* or *being*.

In an article titled "Speaking of Nature: Finding Language that Affirms Kinship with the Natural World," Kimmerer elsewhere explains the link between language and *kinning*. She proposed the pronoun *ki* to identify being and personhood. Rather than use the objectifying pronoun *it* to describe the more-than-human world, *ki* serves as an alternative to animate human relationships with the living world through language. For example, mining companies can't dump toxic mining tailings in *ki* (e.g., the lake).[18] A coincidental parallel here is that *ki* is also the Japanese word for energy or life force, also conceived of as being.

The notion of kinning is a way of verbing the noun *kinship*, animating what it is to be in a web of interconnected relationships. It's less about defining who or what counts as kin and more about shifting language to actively express relationships with all humans and more-than-humans. Being aware and practicing the concept of kinning opens possibilities to transcend separation and ways of treating *ki* like rivers, cattle, forests, or fish as more than objects.

Becoming ecological stems from a language practice that emphasizes the animate nature of actionable verbs that represent living systems. *Earth* or *river* might appear as nouns, but they function as verbs in practice. This is a primary way to enact discourse ecologies in our daily lives – by perceiving language as a vehicle for transforming worldviews through communication and meaning that shifts values, attitudes, and beliefs. In the introduction to the series *Kinship: Belonging to a World of Relations*, Gavin Van Horn makes this proclamation: "Earth – and everything within it, including all that creates what we call Earth – is a verb. All is in motion; all is relating."[19]

> Relating is a fundamental process of *becoming with* the linked web of societal, biological, and cosmic systems that support our existence.

Through an awareness of relating, we understand our deeper connection that moves beyond a subject–object hierarchy. As such, relating functions as an ecological worldview. Kinning is how specific aspects of these systems become more personalized, shaping specific ways of developing these relationships and the connections we may have to them over time.

As Anishinaabe Kyle Powers Whyte described it, kin are like relatives: "relationships [that] can generate high degrees of collective continuance." Kinship extends beyond the notion of relatives as only human and into other species, flora, fauna, and organisms.[20] This is why some countries like New Zealand, Ecuador, Panama, Bolivia, and parts of India have begun to protect rivers, forests, lakes, and mountains by enacting language in their constitutions that acknowledge legal rights defined as personhood for other living organisms in addition to humans.

The recognition of the world's animacy is gradually making its way into legal systems, appearing in constitutions and court rulings, though progress is slow. Kimmerer emphasizes that this isn't just an issue for Indigenous peoples but for all beings on the planet. She claims that developing languages of meaning and understanding, particularly ones that reflect an ecological perspective, is essential, stating that "kinship with the animate" world exists "in every sentence."[21]

PRACTICING

When taking a walk, choose a feature of the environment around you (e.g., bird, stream, or patch of moss). Describe it using verbs to capture its active role (e.g., "streaming" or "mossing") and reflect on how it interacts with its surroundings. Next, reflect on how this element might be *kinning* with you and how its existence supports or connects with your life. Consider doing this with a group and share

> with each other how this shift in language can reshape perspectives on ecological relationships and inspire a deeper sense of care and responsibility in your social groups or local communities. Try to avoid using the pronoun "it." Use "being," "person," or "ki" instead.

Becoming with all life

Synthesizing all the threads of this chapter into one cohesive braid, we conclude by considering the practice of kinning as a fundamental example of relating and do so through a living experience. The article "Living with Bears," written by Ojibway Richard Wagamese, playfully captures the ecological relationship among humans and more-than-humans but does so in a way that features all the aspects outlined in this chapter – decentering the self, relational responsibility, and kinning as ways of seeing language ecologies in practice.[22]

In the mountains of British Columbia, Canada, near his home, Wagamese reflected in this article how bears descend from higher elevations every August, drawn by the abundance of ripe mountain ash berries, rose hips, saskatoons, and wild raspberries. Wagamese focused on coexisting with bears, emphasizing the importance of understanding the concept of "bear time" and respecting a bear's presence in their natural habitat. While Wagamese's Ojibway (Anishinaabe) tradition in Eastern Canada regards bears as protectors with symbolic strength and wisdom, neighbors may perceive them as pests or threats.

The article teaches us the importance of practicing coexistence, even in simple day-to-day tasks like securing garbage and learning to listen during bear encounters. Wagamese views bears as grounding reminders of cultural teachings and ancient wisdom, providing a unique perspective on navigating the changing world and our impact on the environment. Or put another way, bears are one of the many markers of a healthy ecosystem.

Wagamese suggests that bears teach us our place of *belonging* in "a web of life that needs all its parts to sustain itself." *Belonging*, another verb that reverberates across culture as a language ecology, illustrates many qualities of relating. To belong is also to be in relationship to another person, place, organisms, or other species, such as a bear. It also conveys a notion of relational responsibility.

While civilization has placed humans into isolated ways of living, it's a myth to believe that any separation exists between bears and humans. Bears are part of the family to whom we're responsible, to become with them. In this way, bears serve as both a symbol and physical reality.

Enrique Salmón, Rarámuri (Tarahumara) and professor of ethnoecology, coined a term that nicely explains what Wagamese is experiencing. *Kincentric ecology* captures how humans might see themselves as part of the larger ecological family, all connected by shared ancestry and land. Building on the idea of ecocentric (putting Earth systems first), kincentric ecology mixes in relationships that not only involve all living beings but also include mythological or cosmic elements along with practical considerations.[23] *Kin*, as family, and *centric*, as the focus or center, combine smoothly with the interconnectedness of ecological systems, emphasizing relationships at the heart of life.

Returning to the metaphorical and practical story about living with bears, Wagamese portrays a kincentric ecology as a way of relating for both readers and the bears he is coexisting with. The coordinated action between Wagamese and the bear, portrayed with a good bit of playful irony, resembles any other neighbor. There is "bear time" and "bear rules."

Rather than restraints, these distinctions point to the relational aspects of their existence together. The relationship between bears and humans, as depicted by Wagamese, reflects how they "interact." We might just as well say there is "ki time" or "ki rules," drawing on Kimmerer's suggestion toward renovating English language use.

This form of kincentric ecology could also be seen as a language-centered ecology, which focuses on how communication shapes our relationships with other beings like bears. Through Wagamese's story, readers get a glimpse of how knowledge is passed down and shared through different ways of speaking and understanding.

By kinning with the bears, as Wagamese shows, we "see things we might ordinarily miss." In this context, *seeing* isn't just about visual observation but also about understanding the deeper connections between humans and the more-than-humans – like how *listening* or *witnessing* invites us into relationship, *becoming* with our surroundings.[24]

Entangled existence

In many traditions, some of which have been highlighted in this chapter, the notion of a fixed, separate "self" is seen as an illusion, particularly as something the mind constructs but that doesn't seem to hold up under closer examination. Practices within these traditions aim to help people see through this illusion, recognizing that tightly holding on to a personal identity can limit both inner development and outer engagement.[25]

Becoming ecological can support this unlearning of the self. Ecology reveals that nothing is truly independent: a single blade of grass relies on sunlight, water, and soil not just for support to sustain life but as part of what it is. In ecosystems, which is also mirrored in language, interdependence functions as a lived, material reality that challenges the myth of self-contained existence.

The idea of coexisting with all forms of life has been a challenge for humanity for a long time. Humans often prioritize their own needs and perspectives in this diverse web of life across the planet. To remedy this historic tendency, developing relationships again with more-than-human beings through language practices can have numerous advantages. Not only does it enrich human societies, but it also positively impacts all living beings on the planet that interact with these societies. And yet, taking this all in can be a lot to digest.

REFLECTING

- In what ways has the Western concept of the "self" shaped language through cultural values and behaviors that contribute to ecological collapse, and how could alternative views on identity and interconnectedness offer new ways to address this?
- How could thinking about relational responsibility as something we share, instead of as an individual burden, change the way we see our role in tackling environmental issues? How might the way we talk about it help bring people together instead of pointing fingers at others?
- Could developing a *kincentric ecology*, as shown in the story about bears with Ojibway Richard Wagamese, reshape the way you perceive and interact with more-than-human beings in your own neighborhoods or local communities?

CHAPTER FIVE

Ingesting

Taking it all in

This chapter invites you to begin by considering an analogy: just as food nourishes your body, information – especially through language – feeds your mind. Is this more than just a metaphor? The old saying that "you are what you eat" might be true of both physical ingesting and digesting information.

Let's start with the idea that ingestion, whether of food or information, crosses the boundary between body and mind. Food is broken down for nutrients, sustaining the body before being expelled, much like how we absorb and process ideas through language. The body is highly efficient as a processor of nutrients, and it also has mechanisms to dispose of toxins, but traces of what pass through the body remain and accumulate.

What we intake and the way we make meaning with those inputs undergo similar processes, where some parts are absorbed into memory, others stay briefly in short-term recall, and the rest are forgotten or filtered out entirely. The mind-nutrients that are retained can be catalyzed by language into behaviors and habits.

Such a progression presents an infinitely complex system, as is the ingesting and digesting of food, and it's often different for every person. But understanding how a shared resource is consumed and impacts different people or groups in unique ways helps us see how language can drive social renewal and inspire collective action for planetary revitalization.

The previous chapter about *relating* speaks to the importance of *kinning* – a verb that activates our shared planetary responsibilities through language and action. *Ingesting* also has a deep connection to kinship. In many ways, it's the very basis of family units and structures, or at least it used to be, before mass production and the commercialization of food systems.

Just like food, the information we ingest effects our health, particularly our mental health and our disposition toward others. And with the massification of information through modern communications technologies, there's too much to consume. Fasting has always been a way to cleanse the body. Similarly, it's not uncommon for people today to go on an information diet – controlling the intake of various media, news, and kinds of information we consume.

To play with the analogy further, let's ask how much of that information was grown slowly and caringly prepared? How much was microwaved while still in its wrapper and filled with additives that we can't even name?

Oftentimes it feels like we're racing to consume information, with social media reducing our participation in dialogue to a quick "like" or emoji, which is more about reacting to the flavor than truly digesting the substance. Do I like it or not? And if I do, which flavor of emoji best captures my taste?

Most producers of information want others to consume as much of their product as possible because there's financial gain and self-affirmation if they do. The urge to produce information quickly and distribute widely dominates the digital chambers of social media. The result is that, as Paul Virilio wrote in *The Information Bomb*, our sense of time speeds up.[1] We have no time to reflect on and digest all the information that we consume and rely instead on immediate reactions.

This brings us back to the topic at hand, ingesting, which is the most basic aspect of the social order and one of the centerpieces of becoming ecological through elements of language and action. Since prehistoric times, the rhythms of food production, planting, harvesting, hunting, and preserving have shaped how humans mark and understand the passage of time. "In seed time learn, in harvest teach, in winter enjoy," wrote the poet William Blake.[2] This proverbial statement highlights the stages of life and their close connection to another kinning about how we ingest, specifically language.

How does the language of ingesting shape a discourse ecology with implications for the survival of all life on Earth?

Food could be considered a language of the spirit. Eating together is the basis of the social order, not just for humans but also throughout the animal kingdom. There's a particular kind of kinship bond that's forged through sharing food together, taking on many spiritual aspects in various cultures. We also consume information in kinship groupings, and that's why the physical and cultural environment of our upbringing plays such an influential role in our becoming.

For as long as sentient existence, eating food together has been a cornerstone of life. Most religions have communal blessings to say over food before it's consumed. The industrial revolution and developments in transportation facilitated the emigration of many global populations, which then led to the disintegration of intergenerational living practices of families in many parts of the world.

Simultaneous with these social changes, the subjective sense of time is speeding up, from being measured in localized agricultural seasons to globally synchronized atomic clocks. In fact, there's a fascinating relationship between food and time: the more care used in food preparation and eating, the better it is in terms of nutrition, flavor, pleasure, and its ability to strengthen relational bonds among kin. Fast food doesn't nurture us in the same way. This durational aspect of ingesting and health also applies to consuming information.

But it's not just the speed of food production and consumption that's critical to health and pleasure. It's also connected to place, where the food comes from and who is preparing it. There's a collective quality to offering and eating food together. In the process, we're coming together, becoming ecological.

What we also hope to show in the following pages is how deeply ideological food can be as a source of both kinship and conflict, much like the way we ingest language. Competition for food sources has always been fundamental to survival.

And yet there's also a countervailing story, which is the equally dynamic propensity for *mutualism* – a symbiotic relationship where two or more different species interact in a way that's mutually beneficial to both parties. Take, for example, manta rays. Although solitary for much of their time, they gather to eat, mate, and visit cleaning stations, where other organisms remove parasites and dead flesh from their bodies. These organisms are fed while improving the health of the rays.

Species can and do adapt to support one another without either kind being diminished in the process. Another example is pollinators. Those include beings like insects and birds that, through foraging, transfer pollen between plants, enabling their reproduction. This mutual relationship benefits both the flowering plants and the bees while also supporting other creatures, like bears and humans.

The other story of competition does the opposite. The real protagonist of that story is the parasite, which depletes the host until the host can no longer sustain itself. The parasite doesn't exchange energies in reciprocity. Rather, it removes resources from the host and consequently depletes it, whether that's a plant, person, computer, political party, a mountain or the planet.

Humans today have some degree of choice. How we ingest, whether food or information, can reflect either a parasitic or mutualistic mindset. Shifting toward a more reciprocal approach starts with the language we use.

Produce ... fresh food?

To understand contemporary practices supporting ingestion is to grasp the implications of the ideology associated with convenience and then the worldviews they create. As North America climbed out of the Great Depression of the 1930s, a profound shift occurred, one which served the ideology of *convenience* and doing things quickly.

Convenience is marketing terminology derived from Latin meaning "to come together," in agreement and in place, as in the verb *to convene*. Its common colloquial usage today means *made easy*, like *facile* (root of the verb *facilitate*), but without the negative connotations. Something is convenient when it doesn't require much work on the part of the benefactor of the convenience: something that's ready when and where you want it.

In the nineteenth century, by comparison, most people spent a lot of time getting and preparing food, even though industrialization and a move away from local farming were already beginning to change how food was acquired. Food preparation involved complex accounting, community trading, and borrowing as befit current needs and seasonal supplies. There was an entire discursive ecology dedicated to food. It involved complexly nuanced rituals and favors that were sometimes perfunctory, an excuse to socialize, like going to the neighbors to borrow salt but ending up there for tea.

It's noteworthy that the earliest examples of writing from the land of Mesopotamia were clay tablets that represented a certain amount of food supplies that were given by farmers to the collective warehouses under the governance of the king. From this perspective, writing systems evolved partly out of food accounting needs.

Collective eating and drinking experiences to build social bonds can be observed in every culture. Food and drink formed a different kind of currency; it was less about money than divided labor and a way of securing kinship. Feasts were used to mark important occasions in the organizing of social and political bonds, and these events were highly ritualized.

But even in daily life, ingesting was conducted in a ritual manner that wasn't solely about eating. It included all the acts of finding, producing, gathering, transporting, sorting, preparing, and even blessing food. This was all quite time consuming, and some of it involved tiring labor.

The idea of convenience for ingesting was taken up in the United States throughout the twentieth century with spectacular enthusiasm and at a moment in time when there were three particularly aligned forces. The first of these was *mass production*, arguably the most materially successful means of production, popularized by Henry Ford, godfather of the auto industry. Mass production became the de facto solution to everything connected with the American dream because it offered conveniences to every aspect of life, including the procuring and consuming of food and drink.

Another key influence was the birth of *public relations and advertising* because mass producing an ideological stance through influencing what information people consume became about engineering new discourses at the same time as agribusinesses were taking over from family farms. Both kinds of sustenance became massified. And the PR and advertising industries got very busy selling people on benefits of conveniences to improve their lives but with the primary goal of increasing corporate profit.

Closely connected to mass production was the significant change automation brought to agricultural industries. Automation enabled mass food production on a scale that home gardening and kitchen cooking could never achieve. Again, we see a correlation here with the automation of communications, which could spread messages globally at the speed of light. But where mass production saves time and serves a notion of efficiency, automation made it possible to remove people more and more from the production line.

The third force that landed convenience impactfully amidst the food rituals of contemporary humans were the sciences of *chemistry and bioengineering*. These sciences further removed natural processes of producing, selecting or otherwise acquiring, combining, preparing, and preserving "produce" to ingest. English uses the word *produce* to mean fresh food (as opposed to the more congenial "harvest"), like the produce aisle in a grocery story. In this language use, fresh food is a product just like ultra-processed foods (UPF) such as Twinkies or protein bars.

Similarly, recent innovations in computer engineering have made it possible to artificially produce communication through generative artificial intelligence (AI). Because the prospect of machines making meaningful dialogue with us is a relatively new concept, the public is still anxious about AI and its influence on human cultures. The impact, however, has already been demonstrated in the example of automated food. Something essential is lost in the process of not creating our own sustenance, whether meanings or meals.

Mass production of food meant that most people no longer have an intimate knowledge about what they are eating. For example, recipes are no longer shared resources but copyrighted intellectual property – the "secret sauce" that made chain restaurants rich. What's even more inaccessible, the new foods being created for convenience were made using technical processes and scientific language that most regular eaters don't understand. Look at ingredients of a prepackaged food. How many of those words can you say, let alone define?

While food's origins and lingering effects were becoming ever more mysterious to common folk, they were also taking on more ideological potential, as dietary norms no longer had a relationship to land and season. They were more about individualist consumer choice and the language influencing cultural meaning for people that influenced these choices. Every new product was more "mouthwatering," "juicy," "crispy," or "tasty" than the others. Paradoxically, "homemade" taste became a slogan that was used to sell commercial products of mass food production. And taste, like thought, has everything to do with the kinds of discourses we consume and participate in.

With the rise of mass-produced convenient food and drink, a range of new terms entered the dining rituals of North America. Just consider the 1950s with the invention of TV dinners, a frozen complete meal in an aluminum tray that facilitated family eating where the center of attention was broadcast television, rapidly becoming the centerpiece of North American family structure.[3] Dining for pleasure changed in an essential way, becoming less formal about reinforcing connection and more about efficiency, the gustatory accompaniment to the entertainment that media provides. It facilitates individual entertainment, rather than cultivating social bonds.

Consuming media while eating reverses the energetic direction of ingesting, from inward to outward. Food, then, becomes subordinate to some other desire. At the same time, media and advertising promote desires for mass-produced foods and food ideologies, influencing a population that, while consuming and seeking stimulation, is easily drawn into a cycle of manufactured cravings. Food becomes generic and fast, less about the filial traditions and slow, intergenerational knowledge sharing. Advertising directed viewers to mass-produced food and drink, often called "chain" restaurants, that would sell the same mass-produced food regardless of when and where the consumer is located.

Adding to this runaway train, mass-produced food has a secret ingredient: sugared products. Cheap sugar additives, such as high-fructose corn syrup or saccharin, are highly addictive. They provide an initial burst of energy, followed by an energetic crash as the body depletes that energy source. Compare this to environmental messaging that shocks the viewer into sudden emotional responses, followed by subsequent compassion burnout, or the sugar-coating of difficult facts that tout the benefits to economy of ecological destruction. Sugar is the clickbait of gastronomic experience.

The shift in eating habits is deeply ideological,
shaped by the language ecologies embedded
within the social and cultural contexts of
food industrialization.

Changing habits of ingesting influenced shifts in mentality. The institution of eating together, as a collective, started crumbling quickly under the weight of ideology, technological automation, public relations, and advertising, all selling domestic conveniences through broadcast media combined with the increasing mobility of populations that made seasonal food production on an individual level impossible.

This shift led to a situation where people increasingly spent less time engaging with food beyond the act of eating itself. This resulted in less time growing, preparing, or thinking critically about where their food comes from. It became the hallmark of urban living that food production was no longer part of one's social responsibility.

Food preparation was facilitated by prepackaging every possible food stuff, much of it premade, or requiring only one or two ingredients to be added. This phenomenon was also hastened by consumer-end heating and refrigerating technologies, such as microwave ovens, deep freezes, and other kitchen appliances like electric blenders. These inventions further reduced preparation time and the need to preserve food for later domestic consumption.

Whether doing it at restaurants or frequenting convenience stores, eating stopped being about life as much as it was about lifestyle. It shifted from being to doing, from experiencing to expediting meals.

Eating and drinking, by the turn of the twenty-first century, particularly in urban areas, became an expressed form of vigorous consumerism. The daily menu choice gradually replaced the age-old practice of relying on seasonal staples, shifting food consumption from a relationship with seasonal cycles to one of constant selection and convenience. Just think of the sameness of food courts, which are virtually identical in every city, filled with the familiar scent of predictability and the bland comfort of the ordinary.

EXPERIMENTING WITH POSSIBILITIES

What if the word *convenience* had never entered the language landscape of food production? How would your daily routines shift if sourcing, preparing, and sharing meals depended entirely on local community, seasonal ingredients, and intergenerational knowledge? Consider how this would reshape your current relationships, your sense of time, and your connection to the land. Now flip the question: how have words like *fast food*, *instant*, *door dash*, and *grab-and-go* shaped not just your eating habits but also the way you interact with others? Does the language of efficiency outweigh the experience of connection, or might reclaiming slower, more relational food practices offer something richer?

Speaking of eating quickly

Becoming ecological in our thinking, speaking, and acting involves giving the cyclical nature of life more value than the mythical straight line of progress. In the chapter Listening, we looked at ocularcentrism and how a vision-dominant worldview privileges a forward line of sight as the metaphor of choice in language. Built on notions of progress, advances, and innovation, society is moving forward to a not-so-distant future at breakneck speed. If we replace the line with a circle, the urgency to move forward disappears.

This is a listening-centered worldview, one that affects the ways we communicate and think about our beliefs and attitudes. The listener doesn't need to get somewhere else. The more we're still, the better we can hear. The idealized future we dream of isn't some far-off destination. It's already here, now, just unfolding in a different stage of the cycle.

English has evolved to privilege certain expressions over others. As with all languages, communities of users can influence the ways we communicate and act because language is cyclical, feeding back on itself in generative cycles of expression that make culture. This reminds us that the language we use has consequences that transcend the individual, and it places enormous power in collective community-based language change. For instance, calling soil a "living ecosystem" rather than the lifeless term "dirt" signals a deeper recognition of its life-sustaining role.

Cyclical worldviews look and feel different. Who knows, perhaps the pace of a visual-dominated society moving toward an impending future might eventually slow down?

> Ingesting – taking it all in – is no longer just a by-
> product of consuming but a vital part of the cycle of
> life that requires its own space and time.

By becoming ecologically aware, we gain a deeper appreciation for the continuum of what goes into and out of us, and that includes the languages that constantly flow through us. This way, our choices about what we take in and how we process it become a dialogue between the world around us and our inner sense of being.

We can start this dialogue by taking a common expression like "fast food" and exploring what this seemingly innocuous term implies for society at large. This popular food is fast to produce and consume but often complex to digest. The food source needs to be subjected to mechanical preparation and typically is already genetically modified and artificially flavored. Consistency and sameness are of paramount concern if food is to be mass-produced rapidly for commercial consumption.

In the historian David Noble's writing about the social history of automation, he explores the limitations of mechanical automation, and the reason that industries sought consistency in source material. Any experienced chef looks for the very best ingredients to work with, the most fluted tomato to use as a shapely garnish, but that source of creative inspiration can slow or halt an entire production line. Widespread adoption of this kind of automation-ideology has social consequences.[4]

Put differently, fast food is made from something that has had the *personality* stripped out of its genetic code. It's akin to information that has no author or style, more like a fact, unit, or statistic without a story behind it. It's easy to see the connection to generative AI's influence on learning, and its averaging of creative expression. AI is super convenient and saves a lot of time preparing information to be consumed. Fast food is also quickly eaten. Both give the illusion of richness yet leave us with something uniform and less nourishing.

The concept of fast food in word choice and meaning promotes eating fast as an ideal way to enjoy food, particularly on the way to something or somewhere else, elevating eating as efficient and fitting into a modern lifestyle of speed and business. Just look back, not too long ago, to when the drive-through restaurant gained popularity alongside the TV dinner. It's now retro, boutique even, rather than commonplace.

More recently, delivery services like DoorDash, Uber Eats, or Skip the Dishes bring readymade food to the doorstep of urban citizens. And soon, we might be three-dimensionally printing our meals at home, further distancing ourselves from the origins of our food.

Fast food is also reliant on mass production. Mass food typically lacks diversity of flavor. It's engineered to taste, look, feel, and smell the same. Mass food is ultimately about *mass taste* as opposed to refined palettes that tend toward specialized, slow, and labor-intensive production. But this notion of mass food is more than just the food made by automation; it's also the automation of natural processes in plant growth and harvesting or in animal husbandry and slaughter.

Today this ideology of food production underlies eating habits and the social discourses the world over. Convenient food, like convenient information, is produced and consumed quickly. Highly compulsive and addictive, it isn't necessarily healthy.

The most damaging aspect of mass food's takeover isn't just the loss of kinship bonds and rituals that once united communities, though this shift has significantly deepened social divisions and isolated beliefs. It has also promised people constant availability, guaranteeing the food they want any time of the day, week, month, or year.

The problem with this model, even if it sounds good in theory, is that it has altered eating habits. In particular, this has disproportionately impacted poor and marginalized communities who lack the means to afford diverse, nutritious diets, replacing staples with low-quality substitutes. Ingesting has shifted from a dialogue with the environment to a fulfillment of individual desires and tastes driven by mass consumption.

From a language standpoint, the term *mass food* captures this fundamental shift in social eating, from one rooted in local ecologies to one reliant on industrialized technologies supporting mass profit and waste. Mass food production distorts our understanding of where food comes from, especially things like factory-farmed meat, disconnecting us from our directly relationships with food systems.

Before industrialized mass food production, few cultures could afford meat-centric diets (eating meat every day, let alone at every meal). In many cases, cultures would eat meat only on special occasions since raising, hunting, and preserving it wasn't easy or fast. In North America, language helped normalize beef-heavy diets, even where they were once uncommon. From a language ecologies perspective, food terms distance us from uncomfortable truths. Calling cow and pig flesh "beef" or "pork" removes the direct connection to the animal. This allows for eating "veal" without thinking about how it was once a baby cow. By shifting focus from the source to the product, language shapes how we perceive and justify consumption.

As we become more disconnected from knowing where our food comes from and how it's prepared or by whom, we lose touch with the communal aspect of taking in the world, the meaning it creates within us, and the connections it fosters with others. Ingesting isn't a linear process, input to output, goods to waste, to be done as quickly and efficiently as possible.

Humanity clearly needs to process food and information more slowly. Every aspect of the continuum of food systems, from production to consumption, is engineered to make everything happen fast. What seems like a simple expression, fast and mass food conveys an entire ideology of technoscientific enterprise. We're not trying to provide answers to the consequences created by fast-food industries, but this kind of complex inquiry into language is an effective strategy to begin talking about something that is relatable to everyone and affects how we are acting in our daily lives.

Slow food for thought

One doesn't have to pray over food to receive its energetic blessing, but considering what we all ingest, where it comes from, and the quality of care it received is a critical part of becoming ecological. Recognizing that what goes around comes around, as the saying goes, is key. That's what it is to be alive in nested, interdependent systems.

Ecological beings are in dialogue with their environments through what they consume, whether food, air, water, or information. Their environments become a part of them, just as they are part of their environments. Ingesting isn't only about what we consume but also how we use it. What has been lost from the consumerist mindset is that there's more to ingesting than just consumption. It's also about digestion – becoming together – more than input–output or the number of units sold. It involves sustaining life, slowly, and for as long as possible.

Slow food is saturated with the flavors of the environment in which it grows and where and who prepares it. The transference of flavors is a slow process, as when someone marinades uncooked food in a spiced sauce to enhance it. This intermingling of the flavors gets passed on by ingesting and absorbing them.

All the residual traces of the environments of the food, from source to fork, combine, leaving traces in the body of the consumer. This quality of slow food has a strong correlation to the way we use language. When we simply ingest a lot of information and don't digest it thoroughly it can stop being beneficial or nutritious. When language is used hastily, when it's mass manufactured and decontextualized, it can act like a toxin in various discursive ecologies.

The ingesting process in language is about turning inward, chewing an idea over several times, ruminating, like a cow digests grass. It's also about making sense of complex fibers of information. Again, here's another example of how everything is interconnected. How we grow and obtain our food, what we choose to eat, the language we use, and the thoughts we form are all deeply linked. In becoming ecological, we become aware of these relationships and learn to ingest differently.

PRACTICING

Over the course of one or two days, track everything you consume related to food, information, and language. For food, note what you eat, how it was prepared, and the time it took to consume. For information, observe the sources you engage with (e.g., social media, news, conversations, classes, etc.) and how they made you feel or respond. Pay attention to the language used in these interactions. Was it, for instance, nourishing, thoughtful, and relational, or rushed, surface-level, and transactional? At the end of the day or two, reflect on the parallels between input and output. Did prepackaged meals, fast information, or superficial language leave you feeling nourished or depleted? Did intentional,

> locally sourced food, thoughtful information, or relational language help you feel more connected to others and the planet? Share your reflections with others or journal about how these processes influenced you.

What comes out

The chapters of this book are shaped around actionable words, like verbs, and reflect what the Canadian educator Ted Aoki described as "the lived curriculum" – an educational journey of becoming that is always rooted in relationships.[5] Ingesting forges social relations that extend across species and tap into the energetic and spirited core of all being, proving a guide of a living curriculum that can help us become sustainable beings on this planet.

Ingesting is how we become in the world with others. When convenience favors some while creating inconvenience for the majority – whether they're humans, rivers, trees, cows, insects, microorganisms in soil, coral, or wind – we have a clear definition of what's unsustainable. When people talk about the cow as "beef" or coral as "jewelry," the damage is already done because this leads to unsustainable actions.

The language shift of displacing value from the more-than-human to a material object has negative consequences for everyone. As dramatically, the language and information we ingest also has far-reaching consequences. Through language we can influence public consciousness in valuable ways.

A thinking point to end on is to consider how the language of food and eating is connected to the context of food production. Approach this reflexively and see the continuum of nested processes that underly every meal on the journey of becoming ecological. When what we eat no longer completes the vital cycle from soil, water, air, and sun to our plates and back to the soil, it disrupts a fundamental cycle crucial for survival, leaving us all in precarious situations. There's no regenerative, cyclic principle at work. The cycle is broken, and we wind up with toxic waste, so to speak.

Without the slow to mitigate the fast, we lose the ability to relate – to ourselves, our family units, our communities, and the greater context out of which each person emerges. To build generative futures rather than precarious ones, we must attend to the cycle of what goes in and what comes out.

REFLECTING

- In what ways do the sources and preparation of the food or information you consume shape your relationships with others and your environment?
- How does the concept of *mass food* reflect broader social and environmental issues, and in what ways does the language we use to describe food (e.g., "beef" instead of "cow") influence our perception of its origins and impact on the world?
- What practices might help people slow down, do some reflection, and hopefully forge deeper connections through what we ingest?

CHAPTER SIX

Composting

Recycling language

One of the most famous slogans of the environmental movement over the past fifty years is the 3 Rs: *Reduce, Reuse, Recycle*. This has laid a groundwork for many elementary and secondary teachers in environmental education, as well as those educated in this era. People have added to that alliteration since it was coined in the mid-1970s with other Rs of sustainability, such as *Rethink, Refuse, Repair*.

The prefix *re-* in English is derived from Latin and has several meanings, but it generally indicates repetition, restoration, or reversal of an action. *Re-* is also a *phoneme* – the smallest unit of sound in language that conjures cycles and cyclical ways of knowing, being, and doing. It implies doing something again and again, returning to a previous state, or intensifying an action. It conjures continuous cycles. We might then *return* because we may not consider ourselves *responsible* the first time. We could be *reminded* not to take consumption and waste for granted, to *reflect* upon the consequences, and to *reconsider* creating more sustainable conditions that might support shifting attitudes and actions.

In a similar way, efforts to connect language, meaning, and ecology – what we've been referring to as *discursive ecology* – have elevated terms like *remediation* or *remix* to new significance. What's the connection here? Most obviously the idea is to curb impulses that are destructive to the environment and simultaneously destructive to cognition and memory. To discard is to forget. The problem is that when we try to forget all our thoughtless actions, they accumulate because there isn't an *away* to throw them.

Drawing on Susan Strasser's provocative book *Waste and Want: A Social History of Trash*, it might be necessary question the idea of where "away" is when using the common phrase "to throw something away."[1] From a perspective rooted in *place-based consciousness*, or knowing our local environments, we can start asking these types of questions.

Waste has consequences that linger in built (buildings and roads) and non-built environments (forests and mountains). In natural systems, there's no such thing as waste. That's right. NOTHING is wasted in nature. Resources are passed from one cycle and scale of existence to another.

Language, too, is in a constant state of recycling. Words and meanings continuously break apart and come together, rising and fading within the ever-evolving landscape of language. We consider the vital role that these cyclic processes play in invigorating their historical role in life-long learning concerning consumer awareness and waste reduction.

This chapter reconsiders the prominent environmental issue of waste and pollution through the literal and symbolic aspects of composting and the elements of composing language and meanings around waste, reuse, and recycling as cultural concepts. Look at the word *composting*: refuse the *t* and you get *composing*.

Composting is the process of decomposing organic matter, such as food scraps and plant materials, through microbial activity, resulting in nutrient-rich soil amendment. It corresponds with composing, which involves arranging elements to create a coherent whole, whether in music, writing, or art.

> Composting and composing are parallel processes required by living systems and determine the vitality of related ecologies.

The chapter further explores the concept of remix and how we work with memory of other knowledge structures, breaking them down and *recomposing* them in new contexts. We could even say: *remixing* is a form of composting language. To conclude, we will look at how cyclical models of being form the foundation for a larger understanding of *what makes matter actually matter*, highlighting the essential interplay of death and regeneration in all living systems.

Signs in the soil

At this point, most of us are familiar with the term *compost*, which is biological matter like kitchen remains of banana peels or carrot tops that's in the degradation part of its cycle. Compost is closely related to *detritus*, which is the layer of fallen biomatter that covers soil. Just below the visible surface of an organic environment, there's a massive ecosystem of microorganisms dedicated to breaking down organic matter. It's digesting. This makes it possible for vegetative life to absorb the decomposed nutrients through their roots and onward through the cycles of the food chain.

This small and mighty dynamo of life is very efficient. If it weren't, the surface would be constantly rising higher and higher with each passing year, while the soils below would become more and more depleted of nutrients. These small organisms, which vary in size from single-celled life up to earthworms, are known as *detritivores*. Although they account for only a fraction of the mass of Earth's land-born creatures, life would never have evolved without them. They're vital to our survival.

Detritivores also perform another important service for the planet: they aerate the soil. Because most plants can't thrive in anaerobic conditions, this aeration is essential for healthy growth. As detritivores break down the waste layer of biological matter, they release nutrients that plants absorb through their roots and transport to leaves, shoots, flowers, and branches. When animals (including humans) consume these plant parts, they digest the material and extract nutrients. Those who eat other animals continue this process by breaking down flesh and organs to access proteins and other compounds. The remaining fibrous material and fluids are excreted, eventually rejoining the cycle as future fertilizer. It's no different when you eat a salad: your body absorbs what it needs, and the rest returns to the soil to nourish new life.

But not all waste products are the same, with some bearing undigested traces of what has been consumed. Remember that a healthy ecosystem has no waste; there's nothing outside of the regenerative system. Eventually, everything dies and is returned to the ecosystem from which it arose – from soil to soil – thanks to the myriad microcosmic lifeforms that serve the decaying and digesting part of the ecological cycles.

Ecologies of discourse function similarly, especially within the realm of *sign systems* or *semiotics*, which involves the interplay between different signs and meaning that are baked into communication whether they're verbal, physical, or even energetic. Putting this into action, the words "sustainability" or "justice" gain meaning from speech, gestures, and shared assumptions in society, all combining like nutrients cycling through soil. When broken down, language is a synonym for a *sign system* – a way of communicating through established symbols.

Just as soil layers reveal an ecosystem's history, language encodes cultural and social histories through its evolving signs and symbols. In oral cultures, where knowledge is passed down through repeated variations by many voices over generations, information naturally renews itself. This is because no two utterances of a story or even a saying are the same. They differ in intonation, affect, and context of use, with larger variation occurring over time. Each rendition of a story includes variations based on the circumstances and context of the telling.

Over time, new technologies have reshaped these sign systems, changing not just how we communicate but how meaning itself is created and shared. The accessibility of print and digital communications technologies largely eradicated oral human culture. Print changed this natural cycling of information in oral societies, introducing irreversible changes to the discourse environment. Once society became print literate, it was impossible going back to primary oral culture.

Through the presence of visual texts, language turns into a visual and virtual phenomenon, and this changes the way we understand ourselves and the world around us. In oral cultures, to hear voices inside one's head is completely natural. In a print-based culture, it's a sign someone might need help.

In the book *The Breakdown of the Bicameral Mind*, American psychologist Julian Jaynes explores this juncture in evolution, whereby divinities of all different species who had until that time regularly consorted with human beings, rapidly disappeared with the arrival of writing systems. Those who continued to experience the presence of divinities were deemed saints or possessed by demons by religious institutions and, later, as suffering from schizophrenic episodes according to psychological and psychiatric institutions.[2]

In oral cultures, hearing the presence of unseen beings through sound is common. These cultures, deeply connected to their surroundings, understand sound as a way to reveal the hidden energies that shape both the world around us and our inner lives.

Once print arises and brings with it objectivity, language becomes frozen in time. This condition arises when objectivity is the necessary distance between the perceiver and the perceived, the speaker and the audience. For two eyes to focus, there must be a distance between the perceiving subject and the object perceived. Language exits the organic use–reuse cycles and becomes a visual object, alienated from the speaker of the utterances.

Take a book, for example: it's mobile and detached from the author. It's an object, not a subject. Circumstance no longer has any influence on *what* but *how* it's communicated. Consider how political speeches in a media-dominated landscape often prioritize sensation and reaction over substance, using outrage or shocking statements to stir fear or excitement rather than content to educate. Language becomes objectified, rather than personified, as we consider in the chapters on Listening and Relating.

The objectification of language through print-based communications brings us back to the notion of waste. As a prime example, one need only think of the origin of news, particularly the now almost obsolete newspaper. News, as communicated in print as opposed to gossip and the old-world town crier, is a genre that epitomizes waste.

One can sum this up in the derogatory phrase "yesterday's news," meaning it's obsolete or redundant information. An ironic twist is that print newspapers, once dominant, are now nearly extinct, mirroring the shifts from oral to print to digital communication, much like a cycle of use and disposal in composting and composing.

Waste is thus a uniquely human concept. Its entry into contemporary consciousness began with the rise of industrial civilization's concept of mass production and, most notably, with the advent in North America of consumer culture. The opportunity to express increasing wealth and status arose by having new products, what the sociologist Thorsten Veblen called "invidious distinction," or the ability to show off excess wealth through the possession of unnecessary things in consumer-based economies.[3]

Consumer culture thrives on the idea that *replacing* is better than *reusing*, fueled by obsolescence and the latest trends. Look at smartphones: we ditch perfectly good ones for a slightly better camera or a new color like sunset lavender. Another example are all the cats and dogs people adopted during the COVID-19 pandemic, only to give them away again once the pandemic ended. Or, for that matter, people are now more than ever moving through relationships like they're disposable, out of which comes terms like *ghosting* (disappearing from someone's life without telling them). Why bother dealing with a person's idiosyncrasies when many others are out there online? Finding romance online can be just like Amazon, twenty-four-hour service with no commitment to keep it. As we see, human waste eventually transcends objects and animals and moves into people, whether it's through interpersonal relationships or more severe cases like human trafficking.

As humans proliferate and societies grow, mass production is increasingly needed. But, paradoxically, we also need mass decomposition and digestion. The rest of the organic cycle is missing. We might assume that our waste will enter the natural cycles of decomposition. Sadly, this isn't the case.

Modern society seems so confident about this that we have gone to the extremes of one-use nondegradable resources. A well-known by-product of this is the giant islands of plastic floating in the Pacific Ocean, one of which is famously called the Great Pacific Garbage Patch, estimated at 1.6 million square kilometers. As Annie Leonard explains convincingly in *The Story of Stuff*, wasteful, throwaway cultures didn't just happen. They were carefully designed, shaped by clever messaging through language, and are now outdated and harmful.[4] Taken one step further, we might consider the idea put forth by Yale University theater professor Joseph

Roach that "violence is the performance of waste" which "must *spend* things, material objects, blood, environments."[5]

> **PRACTICING**
>
> With a group of people, create a *language compost pile*. Start by sharing a story, quote, or phrase, and retelling it collectively, with each person slightly changing the details, tone, or context. Reflect on how the meaning evolves and adapts with each retelling. Talk about how this process mirrors natural cycles, like composting, and compare it to how written or digital texts fix information in time. Or, on your own, write a short reflection imagining language as a regenerative cycle, like compost. Consider how the stories or learnings you've encountered have changed over time.

Offcycling

Earth's biological systems are entirely dependent on the cycling of matter, or mass taking up space through volume, although this cycling may take microseconds or millennia. This is, after all, a living planet. All life relies on decomposition and reduction to particulate matter. In the reverse of the usual way of looking at things, one could say that life is how the energy made available through decomposition gets distributed. It's also how oral language stays alive and how new words or expressions evolve.

As a move to reorient public attention and reframe these contemporary conversations on waste production, we might substitute the prefix "re" and use "off" instead. Rather than *recycling* (converting waste into new materials) or *upcycling* (converting waste into something of heightened cultural, environmental, mechanical value), *offcycling* permanently removes natural resources from the ecosystem but not from the environment.

This might seem confusing, so here's what we mean: when industrial waste or consumer garbage like plastic litter is disposed of into an ecosystem, it doesn't decompose so that it can be used for regenerative purposes. It takes up space in the environment but isn't part of the cyclical life-supporting systems of the local ecology. As an example, consider how the increasing issue of microplastics in soil disrupt microbial communities, hindering natural decomposition and nutrient cycling that support plant and soil health.

The problem of waste being offcycled is global in scope, though North America wastes more per capita than any other part of the world. Natural resources like water and soil are often wasted, but the problem grows when materials require heavy processing. Mass production of nonbiodegradable materials, such as water bottles, food wrappers, and shopping bags, is driving long-term damage. To put this in perspective, an estimated 250 million plastic bottles are discarded every hour in the United States alone.

Single-use plastics are a stark example of shortsighted, linear logic in manufacturing. As these plastics slowly break down, they enter another phase of existence as nondegradable microplastics and nanoplastics. Not visible to the naked eye, these innumerable plastic fragments are carried in the wind, water, and throughout the Earth's ecosystems, all while being harmfully ingested by all living beings. Microplastics, lodged in blood vessel walls, double the risk of heart attacks and cardiovascular issues, according to the *New England Journal of Medicine*.[6]

Such a linear version of progress is caught in a loop, where the solution to a problem presents an even greater problem. Offcycling is at the center of many of these issues. To get around the carbon dioxide offcycling of fossil fuel engines, industries are creating electronic solutions that present a host of new problems for the environment. Each new technology attempts to fix a problem created by a prior technology and winds up creating yet another problem to solve.

The connection between communication systems and energy supply highlights our reliance on vast amounts of energy and how the non-cycle of offcycling further exacerbates the problem. Keeping the world online 24/7 requires massive energy inputs, not just to run servers but also to power the extensive cooling systems that prevent them from overheating.

The process of manufacturing new materials is also a source of waste. It's not just the end product that's used and discarded. In the process of making new products, many toxic substances, such as paints or acids, are used, resulting in a wide range of toxic by-products of industry needing constant disposal.

This returns us to Strasser's question about throwing trash away: where is this mythical *away*? Is there somewhere we can dump unwanted trash that it won't come back to haunt us later? Unrecyclable waste extracted from natural cycles of composition and decomposition, becoming and decaying, needs a place to be contained and live out its half-life like zombie hordes of undead matter, like *heavy water* (the coolant of nuclear power plants). This problem has been around for a long time.

When we think about bringing language into the process of becoming ecological, the expansive notion of waste can't be ignored. Just as plastic bags and radioactive water persist long after we discard them, harmful ideas like prejudice or denial can linger in our shared discourse, contaminating the cultural soil for generations.

In the sociocultural environments we construct through language, we encounter challenges such as waste, pollution, and toxic by-products, as well as the issue of intellectual property. Intellectual property often removes valuable cultural resources from the common domain, privatizing them and transforming them into objects of wealth rather than nourishing intellectual growth among the populace. These ideas might be packaged for quick sale in the attention economy, but their real value to society has been offcycled.

Mass communication has had many benefits but also unexpected consequences. We're entering an era where digital waste is huge concern. While it's easy to understand how the physical products of technology are detrimental to the planet, we make an imaginative leap when thinking of how the concept of waste affects digital systems because digital waste doesn't appear to take up physical space, just headspace.

From a language perspective, headspace is part of another ecosystem. The astronomical amount of digital information generated and shared daily continues to grow exponentially each year, according to Statista. What's more shocking is that 90 percent of the world's data have been generated in the last two years. Information continues to pile up faster than it can be digested, quicker than it can be composted or even composed into something else. Artificial intelligence (AI) was created to process it all for us, to be the detritivores of digital age, because there's simply too much to process. More and more we rely on AI to upcycle our informational sustenance and to sort out the good from the garbage. In the process, AI creates more information, much more. It is predicted by the time you are reading this, 90 percent of all information online will have been generated by AI. Here, again, there's no *away*, just more information waste that needs even more energy to keep alive.

There is so much content rapidly being produced that we now live in an *attention economy*, where getting people's focus even for three seconds can be converted into mass profit. There's no doubt there are long-term consequences of pouring resources into capturing attention through media – both public and private – especially as it shifts toward energy-draining and computationally complex formats. Let's turn this idea, like turning soil.

Attention is how we take in and ingest information from the world around us, but too much of it can leave us overwhelmed and disconnected. Becoming ecological means focusing on ongoing dialogue, whether with the material or social world, as a way to adapt and take responsibility for the changes happening around us. The language we use for environmental discourse can only inspire meaningful change if it engages in authentic dialogue – among ourselves and with the planet – rather than being set aside, offcycled as objective knowledge and detached facts that leave people feeling powerless in the face of environmental calamities. As we can personally attest from experience, to cite a brief example, restoring a wetland works best when scientists, farmers, and residents collaborate, sharing ideas and adjusting efforts based on the land's needs.

What does this mean in terms of legacies, memory, or collective understanding? The deluge of information isn't dissimilar to what is taking place in the physical world around us. It's heating up because we have produced a great deal of waste in all its formats whether material or digital. We have already offcycled many of the Earth's necessary recyclers, like trees and other plants, faster than they can regenerate. We removed them to make objects, which we then discard and bury after using only once or twice.

But with the explosion of human populations along with their rapidly increasing needs, the hiding places for these discarded objects became fewer and more toxic.

Now even the planet's exosphere, beyond the layer of the atmosphere, is also full of junk like decommissioned satellites and other defunct space technologies. We might wonder if this is a fatal flaw in our genetic code or something embedded in how we learn and communicate through language.

The reciprocity of waste

So far we've looked at how waste isn't "natural." That is, it doesn't exist in ecological systems, where everything is continuously cycled, repurposed, or transformed into new forms of life. Attitudes toward wastefulness are closely aligned with how we are socialized as consumers and woven into contemporary consumerist ideology as a way of thinking that is constructed in language. And waste is just as much an issue in information environments as it is for natural environments.

That's a lot to take in and digest! As well, it's crucial to recognize the bigger picture: how wasteful behavior has become normalized and embedded in our language and communication systems, leading to real-world consequences.

Dialogue, when considering how to build unity through shared meaning, is essential for survival and for understanding ethical practices and behaviors. Almost a century ago, the Russian language theorist Mikhail Bakhtin introduced the idea of *answerability*, which means that when someone speaks to us, reaches out to us in language of any form, there's a collective unspoken responsibility to respond.[7]

Answering makes people accountable to one another, reflecting the cyclical nature of dialogue. True dialogue isn't one-sided; it flows back and forth between participants, and not necessarily evenly, co-creating a continuous exchange of communication that extends beyond just spoken words. Dialogue is also cyclical, breaking ideas down into component parts to regenerate new ones.

This is why we invite all of us, throughout this book, to embrace a *dialogic approach* to what we've been calling our *discourse ecologies*, or the ways language and ecology mirror each other in practice and application. When we acknowledge the answerability of someone's words (also called *utterances* to capture the complexity of unintelligible communication), we're recognizing their place within our shared *we*, allowing their voice to stand alongside ours as part of a collective conversation.

Another way to think about this is to imagine a team meeting at work. When everyone's ideas – no matter how small or incomplete – are acknowledged and built upon, the conversation becomes a collaborative process. Such a scenario builds collective buy-in because it shows that all contributions hold value and nothing is dismissed as irrelevant.

In this manner of speaking, there isn't any waste of language and communication. It all gets used and reused. Dialogue expresses our social connectedness, allowing the collective to adapt to changing conditions in the environment that affect everyone, not just those marginalized or deemed of little value to the social order.

Let's keep exploring the idea of waste from a communication perspective, where it ultimately reflects how we assign value to things. Waste distinguishes all other values because it's the condition of having none. If we don't understand how waste is woven into the way we communicate and make sense of the world, we can't start reimagining a society that aligns with the natural cycles that sustain life.

Defining waste in information systems is inherently complex, as it involves not just excess data but also inefficiencies, redundancies, and the unintended consequences of digital accumulation. Just as the expanding human population on Earth is bringing physical waste to a crisis point for ecological systems, our expanding communicative capacities are generating waste in virtual systems. As the mainstay of contemporary discourse, the rapid densification of the virtual environment is already reshaping social life and impacting planetary ecosystems.

Why are people feeling more disconnected when we're supposedly more connected than ever? What is happening in this ecology of discourse that the systems aren't enabling deeper connections?

As research continues to show, social media directly relates to the immense rise of depression, isolation, and alienation among youth.[8] Not surprising, right? So why isn't language helping people build natural social bonds? Could this be related to the idea of waste within global systems of meaning and value?

When certain voices, signs, or forms of communication are ignored, dismissed, or deemed irrelevant because they don't align with dominant systems of value, they're effectively treated as "waste." A common example of this is when a teenager shares their struggles online, but because it doesn't fit the polished "likable" aesthetic of social media, their words are easily ignored. Their experience and voice get treated as waste in the social media ecosystem of attention and value.

Standing back from these social phenomena and observing them purely from an ecological perspective can provide insights we might not otherwise see. To do so, we need to return to Bakhtin's idea of *answerability*. It's simple but effective: when someone reaches out to us in communication, we're compelled to respond. That's the social contract of dialogue as meaning-making, or relating, and the basis for embracing discourse as ecological. It mirrors the interdependence of thriving living systems.

We have an ethical responsibility baked into us from our very first experiences of making sound to express primal needs. Through this act of sound, we became interconnected persons or beings, and our social language began. This begins a cyclic call and response, a process of living and thriving in dialogue.

But the system is only able to function because it's reciprocal, feeding back into itself as a mutual exchange. When this exchange doesn't take place, we manufacture informational waste, what's littered in the digital media environments and in the airwaves of our social fabric.

In spoken communication, reciprocity occurs by responding, an action propelled by dialogue. A response recognizes the other's personhood and a mutual duty of care. In organic cultures, those shared by more-than-human persons, this call and response pattern is commonplace. Look at birds calling each other, for instance, or even a booming waterfall shaping its rocky bed.

In virtual information environments, the regular currency of the system of dialogue isn't a verbal response but visual attention. Consider how the purpose of clickbait headlines isn't to start a genuine conversation but to grab your visual attention long enough to generate a click, initiating many advertisements. Such a sleight of digital hand turns your engagement upon one click into a currency within the virtual information economy and then wastes that attention once it's received. As a result, people feel both overwhelmed by information that either doesn't enhance personal connectedness or imitates person relations for nefarious means, like phishing attempts to scam users.

As resources for meaningful dialogue shrink, despite an overflow of information, people must compete for attention online, much like they do for food in the physical world. Add to this the constant flood of news updates – minute-by-minute reports about everything happening locally and globally – and we find ourselves in a system that mass-produces information without inviting true dialogue to help consume and digest it. This lack of reciprocity means we're bombarded with data but rarely given the space to process, reflecting and responding meaningfully within these information environments.

As shown in the chapter Ingesting, the cycle remains incomplete without the energy of thoughtful digestion and reciprocal engagement, the circular relationship of input–output, leaving us overwhelmed and feeling isolated rather than being informed or connected.

One-way information flow has a fleeting shelf life, designed for quick attention and single use. Most of it is disposable, lacking accountability or respect for personhood. This creates *virtual waste* that mirrors physical consumerism, where things like production, packaging, consumption, and waste dominate. This one-sided cycle of communication harms both social and environmental systems, leading to effects like vast plastic islands or immense loneliness despite millions of connections.

Returning to our initial question: what makes matter *matter*? Or what drives our endless cycle of consumption through identity, privilege, or the wealth it creates? Once discarded, where do these values go, and what happens to those tasked with reclaiming them at great personal or social cost?

At what point does the enforced end of the cycle, what we've described as the *offcycle*, no longer justify the means or even the meanings? Are we going to be buried by our fickle, ever-changing desire for attention and for new stuff? In a world sinking under the weight of its own material and informational waste, what does true responsibility look like?

EXPERIMENTING WITH POSSIBILITIES

What if everything you acquired, physical or virtual, required you to keep it for a lifetime. That first LEGO toy, mobile phone, or 1998 Honda Civic, all of them no longer working. What if there was no public garbage services or waste disposal centers? What would you do with just the packaging alone, never mind the item itself? The average person in developed countries accumulates a few kilograms of waste every day. How would you cope? If you couldn't get rid of it, if there were no disposal systems or places to throw it away, what would you do?

More than a by-product

Almost like a plausible alternative universe, imagine a scenario where inorganic waste is piling up to the ceiling. You walk through a narrow corridor to go from room to room to find a spot to sit on a chair or lie on a bench of softer garbage. Maybe there's even a single window that looks out on mountains of garbage surrounding you.

The repair shop is immediately the busiest business. Anything labeled disposable, such as single-use plastic or baby diapers, would become obsolete overnight. Arts and crafts to recover the utility of objects, such as making carpets and welcome mats from intricately woven single-use plastic bags, become popular hobbies.

Shopping would become precarious. We would opt to not have our food prepackaged, as was the case 100 years ago. Each piece of packaging or impulsive new purchase would quickly make personal spaces feel cluttered and unmanageable. And every time we acquired something new, it might increase our suffering, not alleviate a desire.

Life, especially in wealthier countries obsessed with many conveniences, would be forever changed. Consumerism would grind to a halt. Industry would similarly be required to be waste-free or keep all by-products on-site making mass production as it's currently practiced impossible. If "away" was our own homes, we'd think and act differently.

In this imagined scenario, we can quickly see how the world as we know it today is built upon the concept of waste and disposal. In many ways, waste is the Achilles' heel of the modern Western project. It's also tied to the concept of the Anthropocene – a period when humans became the dominant force shaping the Earth's environment and future. When there isn't an *away* that our outdated and *off-cycled* stuff or information can be sent to, our lives immediately become unsustainable in a toxic, smelly, disorderly, and disgusting way. Excess and waste would produce a new kind of scarcity: no more livable space.

Our entire system of social values would be forced to adapt. Organic compositing would be seen as godly since we would be relieved of the burden to keep rotting matter. We would love those natural processes that *compose* new life from dead matter or create new works of art from remixing old texts or images or songs or movies we are tired of. We would innovate ways to reduce, reuse, and recycle everything. We would refuse to accept the ideology of refuse.

We close by returning to the interplay of *composing* and *composting*. Composing brings things together, while composting breaks them down, creating the conditions for renewal. This cycling of energy is the fundamental currency of all natural systems. Together, they remind us that growth and transformation rely on cycles of creation, decay, and then re-creation, offering a hopeful image for the work ahead.

By seeing waste not as a necessary by-product of life but as an ideological stance that we have all adopted, coupled with the massively addictive power of consumerism that E.F. Schumacher warned us about in the roundly praised book *Small Is Beautiful: Economics as if People Mattered* way back in 1973,[9] we can adapt language use to start a process of recycling and revitalization of social, cultural, and ecological systems before we trash the whole planet. The deep relationship with waste begins and continues on with how we communicate, even though the topic might be unsettling.

REFLECTING

- How do you think our modern view of waste – both in the natural world and in the way we communicate – shapes the way we live and think? Do you see any connections between how we treat language, information, and the environment in terms of "discarding" or "reusing"?

- How can we rethink or reuse the concept of "away" when it comes to waste, whether it's physical, digital, or intellectual?
- How might you start noticing the "waste" in your own communication? This might be in the way you engage with others personally or digitally, or the information you consume. What small changes could you make to foster more meaningful two-way connections?

CHAPTER SEVEN

Unsettling

Rewilding language

Every chapter in this book compares what's happening in the physical world, such as with global waste, air pollution, systems collapse, rapid extinctions, to a similar social process happening in language. We're not comparing *things*, we're indirectly comparing interwoven *patterns* of behavior in each domain. What we have been learning together is that these progressions often mirror one another.

Language shapes how we think, which in turn influences how we act and relate to the world around us. This raises an essential question: can collective action to sustain the planetary environment that sustains life itself be mobilized through the ways we communicate? To explore this question, we will consider the dynamics of both settling and unsettling in this chapter, examining how language can reinforce familiar patterns or open space for new possibilities.

When we set out to write about unsettling, we knew we were stepping into contested and fraught territory. Why? Because for many centuries, humanity's dominant mindset has been one of conquering the Earth and the many people and life-forms that are part of it. This topic is deeply unsettling because it's about being in discomfort. To unsettle is to disrupt what is fixed, to challenge what feels stable and certain. It's the opposite of being *set*.

From grass to glass and stone to steel, humans have improved their capacity to settle. Settling is about controlling the environment to enhance survival. To settle, and build long-lasting domiciles, brings up another primary function of contemporary social order: the concept of ownership.

The legal legacies regarding property ownership are longstanding, and unsettling them won't be easy or quick. One of the earliest human language texts that has been discovered is the legal Code of Hammurabi, set in stone between 1755–50 BCE, which details the punishments for those who transgress notions of property. The concept of ownership has required a system of laws to support it.

Ownership conveys the privilege of use by one entity of another. It more weakly conveys the responsibility to maintain that entity. The entity that's owned can be anything, but what's essential for ownership to work is that the entity is treated as a thing, a noun-object with a set value. Today, we recognize how offensive and illegal it is to treat people as objects, as seen in the practice of slavery, where a person could be bought and sold at an agreed price. Similarly, land ownership transfers control of entire ecosystems, sometimes not even to an individual but to a corporate legal entity. This concept, tied to habeas corpus, grants legal personhood to institutions rather than living beings.

As we explore how language can spark collective ecological awareness and engagement, we must also rethink ownership and property. How could we imagine alternative ownership models that prioritize ecological preservation on a scale large enough to prevent the collapse of natural systems?

In this *what-if* scenario, human habitation in biodiverse areas like wildlife corridors is prohibited. Society's aim is to rewild areas of the planet to begin to stabilize the loss of species and collapse of ecosystems. Pondering this unsettling idea literally will help to connect this with a practice of *rewilding language* because language not only conveys notions of ownership, but it is also treated as private property.

To rewild language is to diversify it and make it more robust and socially sustainable. To do so, we might consider adapting language to more fluid reasoning, such as acknowledging that terms like "wilderness" carry cultural and historical meanings that shift over time, rather than asserting fixed, one-to-one relationships with what we consider objective reality.

Rewilding language means making space for new and unexpected approaches, ideas, words, and ways of being, and to live in the unknown and uncertain. It calls upon everyone to contemplate the reciprocal relationships we have both with ourselves, our values and beliefs, and with the outside world that transcend concepts of ownership.

With current environmental crises at a tipping point for a while now, unsettling our former habits and habitats becomes an imperative for survival. This type of systemic change always requires unsettling ourselves in a community and society.

To explore how rewilding is a key aspect to unsettling parts of the Earth, we focus on mass migration and dislocation, or the idea of settling in spaces according to the logic of ownership and displacement. And all this leads to the unsettling of language, of course. Hold on tight because this chapter requires a degree of feeling unsettled, of practicing holding space for discomfort while witnessing some of the challenges shaping the planet's future.

We're set, or are we?

Unsettling goes far beyond physical displacement because it disrupts deep-seated beliefs that often include emotional attachments to particular worldviews. The word itself reveals this deeper meaning: to unsettle is literally to remove something from its settled place, whether that place is material, psychological, or cultural.

The word *set* in English historically holds the record for having the greatest number of different meanings, around 430 distinct uses according to the Oxford dictionary. Some of these meanings are profoundly contradictory. The verb *set* means to fix something in place, like using glue, such as when the glue dries it *sets*. As a noun it then takes similar connotations, such as a fixed group of numbers in mathematics, like the *set* of prime numbers. A *set* in the theater is a fixture on the stage providing an artificial environment for the purpose of provoking the audience's imaginations in a particular way. We also use it as a description of character, like when we say someone is "*set* in their ways" and unwilling to change.

But you can also use this word to mean a sudden burst into action, like *setting* off a rocket, or to *set* the volleyball high up close by the net before the spike sends it rapidly descending on the other team's side. If you add prepositions, words like *to, for, on, up, out*, each takes on a particular meaning.

To *set up* implies manipulation of circumstances, as in to set someone up to fail. To *set out* means to move something out of its regular environment. To be *set upon* implies something in the environment that comes quickly and often attacks something else. Or it could mean to *set forth* on an adventure that ultimately leads to *being set* for life, showcasing two contradictory meanings of *set* within the same phrase. Derivatives of the word *set* abound, such as *settle*, when something stops moving, and to *settle down*, meaning to behave in a mature and domestic manner.

So how can we use this word to help *settle* conflict in communication? The contradiction this word embraces means either being stopped, sometimes permanently, or to suddenly burst into vigorous or violent action. It's good to bear this multitude of possible meanings in mind when one contemplates the idea of *unsettling* and how we might unsettle the places where communications get stuck.

Unsettling the mind or our thinking, as we have been saying all along, is equally as prodigious a challenge as unsettling the planet. Interestingly, unsettling suggests both disrupting the old, fixed ways of thinking and doing, while it could also hint at sparking a new, regenerative way of order. When we're unsettled, or experience some form of discomfort, we become seekers, eager to explore new possibilities and meanings with others.

> While failures and errors can leave us feeling deeply unsettled, they're also fertile ground for transformation. In disrupting what we've settled into, they spark unexpected discoveries and remind us that growth like any living ecology depends on disturbance as much as balance.

What is it, though, that unsettles us enough for collective awareness and action to take place? A common cynical viewpoint is that humans are doomed to act selfishly, and that will eventually wipe us out. So why do anything differently? The inevitable outcome will always win in the end.

Humans are, in this pessimist envisioning, utterly predictable. We're creatures of habit that are set in place, and no amount of persuasion and solid reasoning and evidence will change us. We might even see value in the strength of one argument opposing another, like those yearly holiday dinners trying to convince your climate-denying uncle of the "facts" so he might change his mind. For those who choose to take a more hopeful vision into the future, this is a big dilemma, because unless there's abundant and immediate change, that pessimistic vision of the future becomes a self-fulfilling prophecy and foregone conclusion.

Settlement also means to come to an agreement, like when we settle a dispute. To settle any dispute different sides must come together and see other points of view. They must listen and be heard. It requires at least a modicum of sympathetic vibration so that energies align toward a common goal (ending the dispute).

The universe of language is vast, and it should have room for everyone. To settle our differences doesn't mean everyone holds the same opinions or experiences. Rather, it's an exchange through dialogue in which we learn to hold space for other perspectives and for difference, to co-create new meaning as a way of adapting through language.

We need to see more than two sides to every story and to allow the spaces for stories to adapt over time. When something occurs that's disruptive of conventional understanding and political debate, it's much more likely to be disregarded as nonsense, something that doesn't invite serious dialogue.

Unsettling isn't about avoiding difference but looking for and embracing it, looking at it as a starting point to enter into the vast possibility of language and meaning. Find consensus in disagreement, or better yet, question the origin of the disagreement itself.

More specifically, what worldviews or language practices shaped those perspectives? What if before debating a policy idea, we paused to ask how cultural assumptions or unspoken values are already framing what counts as "common sense"? How would that alter how we experience and use language and meaning among multiple possibilities?

Such an unsettling process places us admittedly on the roller coaster called paradox: certainty and uncertainty, comfort and discomfort, or normality or anomaly. But the more we try to feel certain and secure and avoid constant change, the more we separate ourselves from the living systems we're part of. But embracing the irregularities of insecurity and uncertainty is what connects us back to them.

Embracing irregularities

Anomalies, or irregularities in otherwise consistent patterns, appear in many ways. A notable article in *Medium* about AI explains that anomalies like flukes or random noise tend to disappear when outweighed by contradicting data. But looking closer, anomalies aren't just random. They highlight meaningful contradictions and signal urgent complications needing attention.[1] For example, a sudden temperature spike in a stable climate reading could seem only like an irregularity, but it could also indicate accelerating changes.

We're opening with this anecdotal article because it shows how AI processes language in ways that can lead to seemingly incorrect statements, contrasting this with how humans naturally interpret and make sense of anomalies in communication. To this point, language that's unpredictably wild and "sensational" gets our attention, and we might think it's done to trick us. Just think about how people react to irregular speech, shifting from curiosity to surprise or even shock.

Although there are an infinite number of possible expressions in English, people are also predictable enough for a statistical algorithm with generative AI to both imitate and perfect these expressions. This kind of habitual language use creates a predictable mind*set* or worldview that's difficult to unsettle. Put another way: we encounter anomalies when our expectations or predictions don't match an accepted reality.

Think of the platypus – an egg-laying, duck-billed and -footed venomous mammal with the body of an otter and the tail of a beaver. They don't have stomachs or teeth, and underwater they see with their beak, using electro- and mechanoreceptors. Like many anomalies, they're endangered, another species at risk of extinction. Biological anomalies aren't always natural in origin. Sometimes it's a result of toxins introduced to their environments, such as two-headed fish that show up in the Great Lakes of North America.

Another anomaly that warrants attention isn't so much a difference in *kind* but in *quality*. According to the Austrian engineer Victor Schauberger, water has something called an anomaly point. At 4 degrees Celsius above freezing, water takes on very special properties. Any colder and it begins to change state and freeze into an expanding solid. Any warmer and it loses energetic properties. But at 4 degrees water has unique and somewhat mysterious properties, particularly in terms of energy.[2] As it turns out, this is also the temperature of water when it comes out of the mouth of high mountain springs in its purest condition.

Reflect for a moment. Mountain spring water rises through rock from hundreds of meters below. It defies gravity. Rather than being pushed up by pressure from below, it's drawn to the surface of its own volition. That cohesiveness, sometimes referred to as surface tension, provides the kind of energy that enables objects that are heavier than water to float on its surface or for drops of water to remain suspended, say on a leaf during rain, when by all scientific accounts they ought to succumb to gravity like everything else. This is unique to water, the irreplaceable foundation of all life on Earth.

You may be wondering what this odd information about the anomaly point of water means in terms of language? For starters, water is itself a language communicating to all forms of life on the planet. The author and social commentator Ivan Illich reminds us that water can "speak and sing in seventeen different registers."[3]

Looking at this from a different angle, language is like water to the ecosystem of thought. At the right temperatures, all manner of discourses thrive, bringing us together so that we may accomplish seemingly impossible tasks. When discourse is heated by anger or resentment, it can do the opposite. Language becomes destructive. At its most potent is when it's drawn up from the depths of the psyche, in those reflexive caverns of neurons, and rises to the surface springing into verbal or written expression, just like water.

Beyond words exchanged on the surface, language
runs through us like water, becoming thought,
becoming the stories that guide
our lives.

But all anomalies are significant because they demonstrate a difference that defies categorization. They aren't *set* and so don't *fit the mold*. This means humans, like AI, need to resolve the contradiction they present to an ordered worldview. And without a doubt, the current world is presented with all kinds of anomalous data, such as snow falling in Cuba or polar bears living without ice.

While the connection between water and anomalies may seem unrelated at first, we use this ecological analogy to highlight how the cyclical nature of language mirrors the dynamics of water cycles. Patterns within natural systems are indicative of larger trends often mirrored in language. This is where the scientific concept of a *tipping point* comes from: a moment when the pattern, or model of the behavior of some phenomena, changes so significantly that the model will no longer serve predictions.

One of the most common references to these tipping points comes from climate science, particularly through the 1.5+ degrees rise in global temperatures that we hear so much about. Reaching beyond this degree of overall planetary warming, which is already happening, pushes us into an anomalous situation, where previous models of prediction are longer be applicable.

We enter uncharted territory, and likely we're already at this tipping point. Anomalies in the normal operation of systems are becoming apparent. What's more, they're increasing all the time. How does language facilitate our appreciation of systems that have become unpredictable?

News about ecological collapse is full of instances where new records are being *set*, and simultaneously this news is psychologically unsettling and causing a significant rise in anxiety for people. Norms relate to expectations and are a basic aspect of our psychology. Norms tend to be intergenerational and often passed on from parents to children and teachers to students. Norms give us a sense of security in the ability of patterns to remain constant over time.

One key area where we witness this kind of disruption involves environmental discourse. Even among allies there can be imbalances in power that lead to stalemates lacking in action or even opposition. The progressive viewpoint on equity, diversity, and inclusion highlights the need to emphasize the voices of people who have been historically marginalized based on, among other things, varying levels of systemic oppression. As we become ecological, we start to recognize this kind of exclusion and inequity may start with humans but then goes well beyond interactions with our own species. It speaks broadly to our relations to the Earth and the ways in which we collectivize care and awareness through dialogue.

To be radically open to the future is to also be truly open to anomalous change, unexpected dialogue, revealing contradictions or irregularities within a system, and to welcome the exception from the norm as a catalyst of change. We're suggesting that the energy needed to catalyze social and behavioral change can come from language and is released in the change of states, from being settled to becoming unsettled, like when ice, the fixed state of water, melts and flows again. This is how language and meaning evolve, shaping social discourse as we learn to speak and act with ecological awareness.

> **EXPERIMENTING WITH POSSIBILITIES**
>
> What happens when the world's anomalies like heat waves, melting icecaps, or mutated species challenge the very models we use to predict and control? As scientists look for patterns, what if the patterns we rely on no longer fit? Water has an anomaly point at 4 degrees Celsius, where its properties change radically. Could this be a mirror for how language and meaning evolve? What's the anomaly point of political discourse? If environmental irregularities are increasingly disrupting natural systems, do we also need to reconsider our assumptions? What if the data we trust aren't enough anymore and we're facing a shift into uncharted territory? How can we adapt when old models no longer serve us, and how does language help us navigate this change?

To be or not *to be*

To unsettle ourselves might be unnerving, but it's also the impetus for change. Many of us wind up with habits and behaviors that we have settled into over time. Unsettling these comfortable patterns take lots of effort while also bringing endless adventure. As the professor of consciousness Donna Haraway writes, real change requires us to "stay with the trouble" – to remain present in complexity rather than rush to resolve it through partial solutions, like taking sides to quiet the dissonance.[4] Another way of saying this is sitting in discomfort or unsettling in uncertainty.

The challenge of unsettling a language is equally as prodigious as unsettling a land. Postmodern philosophy in the mid to later twentieth century, including movements such as *deconstructionism*, attempted to do just that: developing a *discourse* that challenges traditional language norms and expose their influence and power in society.

Thinkers like Roland Barthes proclaimed the death of the author as a concept.[5] This idea connects to everything we've been talking about in this chapter because without an "author," there can be no intellectual property, no ownership of language. Meaning-making shifts continuously as a common resource like it does in spoken language.

A fellow French philosopher, Jacques Derrida, pointed to the linguistic root of our settled preconceptions in language, particularly through the verb *to be*.[6] Derrida was focused on debunking the mystical elements embedded in language, and so it's worth teasing out his idea in the context of unsettling.

When we make a statement about reality, we often use the verb *to be* like an = (equals) sign, as if we're fixing something in place. For example, when we say "the climate *is* heating up," the verb *is* ties our idea to something we treat as an unquestionable Truth. But Derrida argued that there's no single, essential truth, just shifting meanings. In other words, the *to be* verb (called the *copula* in linguistics) doesn't prove reality; it only helps us imagine it, not *set* it as fixed. For example, saying "this forest *is* healthy" depends on who's defining "healthy" and why.

What we often think of as reality or outside of what's written is actually just another kind of text. In this sense, a "text" isn't just words on a page. It can be anything that we interpret to make meaning, like a conversation, film, building, or even a gesture. Meaning lives not so much in the object itself but in the web of relations we weave around it and among ourselves. That's why understanding is something we build together, as in relating, often by talking about things through the process of dialogue. Returning back, it's also why the concept of the author is much more than a single person but engagement with various *texts* in environmental *contexts*.

There's more. In environmental debates, for instance, people try to assert a fixed reality. As often the case, we attempt to settle disputes according to our own perception by using this verb *to be*. Here are two simple examples of how this could play out: the Earth *is* warming; climate change *is* a natural cycle. We mark off ideological turf with the *to be* verb *is* and compete for dominance. While in heated debates, we resort to this verb more than any other. The verb *is* holds our claim to knowledge.

And the more one version of *to be* is repeated, the more *settled* it becomes in our thinking. A lot of self-help techniques in books or on TikTok involve positive reinforcement through repetitive use of this verb: I *am* strong. I *am* capable. I *am* loved, and so on. This transformative act of repeated language has a history of use going back to the ancient oral practices of chanting or reciting mantras.

On the flipside, this kind of logic leads to generalized assumptions: all environmentalists *are* kooks, technology *is* destroying the world, this person *is* wrong, and so on. These notions get entrenched through repetition in the dominant narratives that construct our worldviews and bias our understanding.

They're hard to destabilize, or unsettle, even, in the face of contradictory evidence. That's exactly Derrida's point. Reducing truth to *linguisticism* – the idea that language contains fixed, essential truths – overlooks the ever-changing, diverse, and unpredictable realities that go beyond the patterns we label as truth.

Understandably, we might think that language that lacks this force seems wishy-washy. A statement like "climate *could* be heating up" means it's open for debate, not an open-and-shut case. When we want to say something important, we resort to making truth claims about it, and the battle begins: "climate *is* heating up" versus "global warming *is* a hoax." To back up truth claims, we then, rather predictably, resort to possible truths relying on the *to be* verb, the *copula*.

> To unsettle language and make new habits of thought and behavior possible, we might need to start with something as simple as how we use the verb *to be*.

At its core, the verb *to be* asserts a fixed identity to something. To avoid being outside the norm, we adopt ideological positions that conform to an identity type and hold onto them, even when they become out of touch with the needs of our times. We need to make new connections, more fluid identities and self-definitions, which is only possible if substantive change in human behavior is to be activated.

As we consider in the following two chapters, Storying and Witnessing, narratives and testimonials move us by linking knowledge to lived experiences rather than objective fact and open us up to the possibility of unsettling people's subjective positioning.

Replacing the verb *to be* with more descriptive verbs takes work. This is because it shifts the nature of dialogue, making it circular rather than bipolar, as in opinion A versus B. We might begin with sharing circles instead of debates, removing the personal subject "I" at the beginning of every sentence, or refraining from *to be* verbs when making controversial statements, leaving room for possibility. In doing so, we open ourselves up through the act of listening to various stories and engage in healing circles rather than argument.

No one owns the Earth

To finish off here, let's consider Edward O. Wilson's Half-Earth theory. Wilson, a Pulitzer Prize–winning author and ecologist, proposed that dedicating half of the planet's land and oceans to protected natural habitats would help preserve biodiversity and prevent mass extinction of species. His idea was that land animals need to migrate, and human ownership and control of land make that impossible.[7] Each incursion of humans into ecospheres weakens their resilience because we do so under a concept of owner-ship, which prioritizes the human individual over the entire cyclical systems of a place.

By claiming ownership, we cut off the woven stories of places and erase those of the more-than-human world. To destroy is also to *de-story*. Wherever humans go, we make the natural environment inhospitable to any other species, except for a few favored varietals like dogs and cats. To pre-vent massive extinctions of living systems, Wilson's Half-Earth theory may be our Hail Mary, the one last fleeting change to reverse the damage.

The human narrative that's written on place uses indel-ible ink, so to speak. It's almost impossible to erase. And to remove it, we would have to radically alter histories of human governance and power. It might seem unlikely, at least at this point in history, but it's certainly possible. There are historical examples where tides have turned for the bet-ter, such as the abolition of slavery, women's voting rights, civil rights and anti-apartheid movements, efforts to limit colonial expansion, and, in an ecological context, the reduc-tion of the hole in the ozone layer.

Let's not think of ownership and control as inevitable and absolute. Even though there are very few places exempt from human domination, this doesn't have to remain the norm. Humanity may currently own the Earth as its property, but this notion can be unsettled through both language and action.

What would it take to make the entire human species
unsettle and migrate in its thinking toward something
like the Half-Earth model?

The difficulty of this ask might seem enormous, but what we're suggesting is that some conventional practices of language are limiting those possibilities. Dare to imagine as a starting point that changing the way people use language begins to shift the storied beliefs that have allowed for these conditions to emerge in the first place. Yes, imagination can drive these changes. Language has the power to conjure the impossible, as the poet William Blake reminded us: "What is now proved was once, only imagin'd … One thought, fills immensity."[8]

When we talk about rewilding language, it's to promote language that's complex and alive, making it diverse and adaptive like any other ecosystem. Language sustains and nourishes our minds as physical ecosystems sustain and nourish our bodies. Rewilding language allows ideas and thoughts to diversify and to exist in interrelations of mutualism, not competition.

Language can help us shift toward localizing and reusing, ending waste, and ensuring that extraction is followed by repair and natural regeneration. The Earth, now smaller due to human growth and mobility, can no longer sustain us if we continue with a mentality of ownership and dominion over other species and living organisms. It's time to unsettle both our language and the planet.

Now, let's grab bill bissett's poetry book *Nobody Owns the Earth* and start reading, reminded that our ideas of ownership and control – whether of language, land, or ecology – won't lead to a livable planet.[9] Adapting to this new model is uncomfortably unsettling. What a radical concept for humanity to comprehend: no one owns the Earth, but everyone is responsible for it!

This unsettling process opens up chances for change by shaking up dominant worldviews that are overdue for revision, and what comes next is learning to restory them.

PRACTICING

By yourself or with others, reflect on the idea of land ownership and migration. Begin by exploring what "home" means to you. Next ask: what would it feel like to be displaced, whether by a weather event, politics, or migration of species? Then, consider the concept of "uninvited settlers," whether you may have experienced this or not, and how human history has shaped the ways people think about land and ownership. How does the idea of *settling* and accumulating possessions (on land) compare with the rights of other species or Indigenous communities to move freely or protect their land? Using E.O. Wilson's Half-Earth theory as a starting point, talk about how we might reimagine the use of the planet. What would it mean to allow wildlife to move freely and live according to their needs? How could we act collectively to protect biodiversity while balancing human needs? Finish by sharing ideas on how to apply these concepts to everyday actions and decisions.

REFLECTING

- In what ways can experiences of being unsettled, whether through language, differing perspectives, or disruption of familiar ways of living, serve as catalysts for deeper understanding and collective transformation?

- How can language encourage us to embrace anomalies, the irregular patterns, as opportunities for change rather than threats to stability?
- How does language shape our understanding of place and ownership? How might we rewild language to shift our practices of *destorying* place to *restorying* ecologies of place?

CHAPTER EIGHT

Storying

We're all stories

Pause for a moment and consider the stories that have shaped your views toward the natural world, species other than humans, or suffering caused by extreme weather events. What kind of stories were they? Are they from your childhood? From books? Or, perhaps, from TV or podcasts? Are they positive? Scary? Nostalgic?

Now, choose one of these stories and consider for moment why you believe it had such an impact on you. How did it change your beliefs or behaviors?

Although storytelling may seem too subjective to affect society's response to ecological collapse, it nevertheless plays a significant role. One well-known environmental example is the "Fridays for Future" movement, led by the now famous Greta Thunberg, but there are hundreds of others also doing this type of work. By using storytelling techniques to amplify youth voices and personal narratives, it continues to inspire climate action worldwide.

Author and professor Thomas King explores the universality of story in our social histories and relationships by affirming the "truth about stories is that's all we are."[1] Stories have the power to shape and also heal our lives. Novelist Ben Okri also recognizes our intrinsic connection to stories: "We are part human, part stories."[2] This interdependence goes beyond prevailing theories of individualism, emphasizing the deep magnitude stories play in our existence and how much that can affect change.

The truth is that stories are everywhere, which is why they're also deeply connected to place. The world is composed of a multitude of stories that shape societies, communities, families, and people, serving as the foundation of human existence. It's no surprise, then, that the author Salmon Rushdie calls humans *storytelling animals*.[3]

On our own and as part of a broader community, peoples and cultures around the world rely on images and metaphors through stories to shape values and aspirations. Storytelling, rooted in personal and social narratives, is fundamental to being in the world. It's astonishing when stories are often dismissed as merely subjective and seen as less credible than scientific facts or technical evidence, which are often stripped of context, removing the heart and soul of experience.

"The universe is made of stories," proclaimed the poet Muriel Rukeyser, "not of atoms."[4] This cosmological invitation encourages us to view stories not just as narratives but also as a fundamental language, akin to the communication of sub-atomic particles throughout the universe. Language derives its meanings and understandings from a blend of collective and personal stories, both conceptual and tangible. Both are composed knowledges created to explain our existence.

Stories help us make sense of any situation. To describe an elaborate experience, we tell a story. To tell our friends about a blind date, we tell a story. To capture a life-changing event like an accident, we tell a story. To convey the importance of a ground-breaking scientific discovery to the wider public, we tell a story.

Stories rooted in a place or location also evoke some form of memory or emotion, like where we first fell in love with the natural world on a trail somewhere in the foothills or wooded forests. Stories bring settings to life in the imagination.

Imagine a narrative set in an old-growth forest full of biodiversity, where a community of locals in the region have lived for centuries. As the story unfolds, we learn how the ecosystem has been disrupted over time by the encroachment of industrialized logging companies sacrificing old-growth forests for profit. Through descriptions and character interactions, the narrative illustrates the interconnectedness of all life-forms within the old-growth forest and the immense impact deforestation has had on both the environment and the community.

This simple universalized story is meant for accessible appeal. This is just one *possible* story that resonates for both of us as inhabitants of part of the North American west coast with a deep history of old-growth forests and their ongoing disappearance through logging for industrial growth and profit. In this example, storying serves as a powerful tool for conveying the interconnected complexities of an ecological issue in a way that resonates with people's lived experiences.

This chapter explores how the ways people use stories can make or break relationships with our environments. It also looks at how language possesses an uncanny ability to catalyze change in perceptions, beliefs, and values through the process of storying. Whether it's reflecting on collective climate stories or considering how societies and cultures write and rewrite stories about their relationship with the environment, *storying* – as an actionable way of using the language and meaning – plays a significant role in how to approach environmental communication and what meaning comes out of it.

> Because storying is circular and continuous,
> it functions as a process of relating, where
> meaning is made between or because of the
> participants, not just as an individual or personal
> accounts on their own.

Storying is also part of discourse because it involves the exchange of narratives within specific contexts in place and time. When people or groups share stories, they participate in shaping meaning, especially in how it's constructed, negotiated, and communicated through languages and narratives. Storying teases out meaning within a given cultural or social context. Different interpretations of the same story can lead to exchanges of ideas and experiences, contributing to the ongoing construction of shared meanings and identities. It can also assist us with the language and emotional tools to deal with being unsettled.

Providing the capacity to clarify our knowledge and values, storying allows us to reflect on our lives and, consequently, forge stronger connections with the ecological settings that surround us. Storying is a reciprocal feedback loop that both responds to and builds from our experiences to change our lives and societies. Just as we make up our own stories and ideas about our social lives, we also shape how we see and interact with the environments around us.

Finding the plot

We purposely lean into the verb *storying*, a way of including but also adding to the terms *stories*, *storytelling*, and *narratives*. Although somewhat distinct, they're all closely connected. Stories are made up of events with things like characters and settings that form a larger narrative. In simple terms, storytellers share stories, and these stories come together to form a broader, more complex structure known as a narrative. Storytelling is the process of sharing these stories as part of that larger narrative. This overall practice is more than just an act of doing; it's an interrelated web of storying.

Many people use stories, going back to Aristotle's definition in *Poetics*, because they have a plot and refer to specific real-life accounts or tales about people, including what often happens to them in a course of time or period. Stories are more familiar than narratives because they have a predictable temporal structure of a beginning, middle, and end. They also contain identifying literary devices, such as themes, characters, settings, plot, conflicts, point of view, and resolution.

Narratives are more complex, with broader accounts of events, experiences, or happenings. They provide patterns and structures that help to make sense of stories. They serve as more open-ended and participatory stories, with various interlocking elements of layered stories that contribute to various understandings or systems of thought.[5] If the story is a captured moment, then think of narrative as the framework organizing and making sense of that moment.

For instance, narratives can explain more about worldviews through specific methods of relating to the world through *master narratives* (dominant worldview), *metanarratives* (shared across subgroups), or *counternarratives* (alternatives to master narratives).[6] Narratives also offer a space for nonlinear and fragmented exploration, much like expressionist film or abstract art, which challenges certainty and invites embracing the messy nature of existence.

While narratives may encompass a broader range of accounts, and stories might consist of specific tales with identifiable elements like characters, plot, and resolution, storying includes elements of both but extends beyond them in the ways they construct meaning through language. Storying is the practice of living with and through stories, as an ongoing weaving of meaning in dialogue, where knowledge, identity, and even our sense of self are remade in the currents of interaction.[7]

We acknowledge that definitions can be confining, so these are mere placeholders to build off. This is why the term *storying* is used here as an integrated model of doing, being, and becoming stories as participants, listeners, witnesses, and tellers of the systems of language and meaning-making that function through stories but in larger narratives. As many local and global cultures have known for millennia, storying is a more relational and holistic way to convey information or learnings.

Even today, narrative stories are transferred through streaming TV, podcasts, or films, whether these are master narratives or counternarratives or purely fictional imagining. Think of an obvious example that we've all been exposed to: documentaries and films to raise awareness about pressing environmental issues and inspire action. Over the past few decades, documentaries like *An Inconvenient*

Truth by Al Gore, *Our Planet* narrated by Sir David Attenborough, and Rebecca and Josh Tickell's *Kiss the Ground* have effectively highlighted the greater consciousness around topics like global warming, habitat destruction, and species extinction. Despite tendencies toward overdramatic scenes or proposing easy solutions, they communicate to broad audiences, serving as catalysts for public dialogue leading to advocacy efforts and community action.

Reframing our ecological narratives is one way that draws on using language and the actions that emerge from it to change possible futures through stories. Stories hold knowledge and information, but what makes them produce meaning as language ecologies are the ways they function as multiple sites of meaning. They're used to work through various language forms and levels of speech, as well as cultural dynamics. Stories help to organize human structures and development, providing favorable conditions for becoming ecological.

Creating worlds

Serving elements of language, words matter because they create meaning in our shared worlds. In addition to creating meaning, words create vast possibility through multiple worldviews. This meaning establishes the conditions for societal stagnation or renewal. Even historically, the way human infants learned language in the first place was by hearing and reading stories.

The language of stories can both heal and alter our lives. They are also the lifeblood of our language ecologies. As noted by the nature writer Barry Lopez, "everything is held together with stories."[8] They bind us, even when we might not see it, as connective tissues of existence.

Nicholas Money, who is a mycologist, made some compelling discoveries about how fungal networks not only communicate with each other but also how they can sense one another, learn from each other, and make decisions. Mycelia, which are fungal threads that look like branches, are capable of spatial recognition and might even have short-term memory.[9] These fungal languages remind us that communication is not the exclusive domain of humans; it is a property of life itself. Like trees exchanging nutrients through underground networks or people shaping cultures through narrative, all systems, whether ecological or linguistic, are collaborative forces. They build worlds together, telling stories in languages we are only beginning to learn how to understand.

Language about storying the future plays a crucial role in shaping policies and actions related to climate change, influencing how individuals and societies perceive and respond to this existential global reality. One example of this might to be to look at how communities are increasingly recognizing the need to build resilience and adapt to the impacts of changing weather patterns due to climate shifts.

What's interesting, though, is how the language keeps changing. Terms like "climate resilience," "community-based adaptation," and "climate justice" are gaining traction as communities collaborate to address vulnerabilities and enhance adaptive capacity. Each of these terms adds to the larger narratives that story our lives. They will also adapt and change over time, along with the cultural and linguistic shifts.

Such terms like "natural disasters," which disconnect human responsibility, are now called "extreme weather events" or "climate-related disasters." "Climate denial" has been rewritten as "climate disinformation," showing the active role that narratives play in ignoring credible evidence. Even "global warming" or "climate change," as we've mentioned, has been reconsidered as "climate breakdown" or "climate emergency," which speak more to the overall experience and reality of "ecological collapse" and "mass extinction." Climate is no longer treated as a science so much as it is a euphemism for emergency or urgency.

Rather than remain limited to terms like "warming" or "heating," this terminology captures the destabilization of atmospheric changes that affect all weather systems, resulting in extreme events, such as floods, droughts, freezes, hurricanes, and so forth, that massively affect large global populations and societal infrastructures. Through storied accounts about these events, even people who didn't witness them can start to relate to the experience of those extreme climatic events.

Another example could be seen in how increasing recognition of the importance of regenerative agriculture practices that enhance soil health while sequestering carbon from the atmosphere. Words like "carbon farming," "soil regeneration," and "agroecology" are increasingly prevalent vehicles for stories, as farmers and policymakers embrace holistic approaches to land management.

Even compounded words such as "permaculture," combining "permanent agriculture" and "permanent culture" to convey its dual focus on sustainable food production and social structures, are now common language in ecological discourses. As new words and expressions are introduced to contemporary language practices, new elements are also being added to storying practices. What are the stories that are told to support "extreme weather events" or "ecological collapse"? These are changeable placeholders in our language histories, but the stories themselves are what mark these moments and give them meaning.

Co-creating different worlds relies on shifting language and action through an active process of storying.

While this process can be overt in films, TV, pop music, or novels, it also works more subtly, shaping cultural discourse over time through everyday language shifts, as seen in the earlier examples of terminology. These less explicit cultural and linguistic storytelling practices might wield even more power and influence than larger, emotionally impactful examples, as they're deeply integrated and interconnected with cultures and communities.

Less conspicuous forms of communication, such as imaginary or metaphoric language, also exist and influence our sense of interconnectedness with the world. Humans live through *metaphor*, which is figure of speech that describes one thing by referring to something else, showing a connection between the two. Language is a web of conceptual metaphors that have the power to mobilize people through their capacity to affect meaning. Even if we're not consciously aware of it, we largely think and speak in metaphor, and our stories unfold in our minds like a continuous movie reel.[10]

Metaphors, as a key element of storying, play a symbolic role in how we convey meaning through images and ideas. Take the metaphor "war on climate change," for example. It's often used to highlight the seriousness of the issue. While this metaphor can be counterproductive sometimes by associating climate action with violence and oppression, it does flag for us a sense of urgency.

The metaphor of war can also apply to the battle over information, where controlling the narrative through dis- and misinformation shapes how people perceive the "right story." Elections are now "won" by overwhelming the public with any type of information, regardless of its truth. As discussed in the chapter Composting, these *information bombs* are forms of waste, both of language and energy. But these conditions also arise from the metaphorical nature of storying.

As we've been noting here, one complication in contemporary society is the tendency to overlook the metaphoric foundation of existence and interaction. This brings to mind Apalech Tyson Yunkaporta's sage words in his book *Sand Talk: How Indigenous Thinking Can Save the World*, where he claims that "metaphors are the language of spirit," which we might also interpret as the language of the more-than-human world. In this same vein, systems scientist Gregory Bateson called metaphors "the logic of the organic world."[11] Yunkaporta takes this idea further, emphasizing that "powerful metaphors create the frameworks for powerful transformation processes."[12]

What's particularly useful about Yunkaporta's teachings is that they illustrate how metaphors can be a generative way to understand and convey stories. In particular, they can evoke emotional responses and stimulate empathy by connecting environmental issues to human experiences and values. Symbolic language resonates profoundly through storying, and recognizing metaphors in each story opens numerous possibilities of meaning and future-making, contrasting with the limitations of one-dimensional literal interpretations.

But people must also choose metaphors carefully, considering their potential impacts on public perceptions and policy decisions. The metaphor of a "carbon footprint," for instance, makes climate impact feel personal and encourages individual responsibility, and it can also distract from the larger systemic changes needed at corporate and policy levels. Understanding the underlying metaphors in ecological discourse can provide insights into cultural attitudes and values, as well as power dynamics. The choice of metaphors that we take up in our stories can have direct social and behavioral outcomes.

Let's take the language of storying one step deeper. *Metonymy* is a cognitive process where one entity is used to represent another, revealing conceptual structures that shape our understanding of the world. In other words, when we say "the land is speaking," we use the whole ecosystem to represent the many organisms within it, showing how our minds understand through layered relationships.

This process goes beyond mere reference; it also helps us comprehend abstract multidimensional concepts by associating them with more concrete ones.[13] Much like word choice and metaphor, metonymy is another language tool that can affect people's responses and awareness to their ecological relationships.

When researchers talk about "saving the bees," they're using the term "bees" metonymically to refer to the broader ecosystem and biodiversity. Conversations about "protecting the oceans" often involve using "oceans" as a metonym for marine life and environmental health overall. And perhaps most famously, anything related to the term "climate" isn't often about climate science anymore, which has been quite established and consistent for decades, but more about social awareness and action (or resistance in some circles). Climate has become, for now at least, the de facto metonym for planetary health and primarily human and more-than-human survival.

Metonymic highlighting can show how dominant ideas resonate with groups, shaping their beliefs and actions. When policymakers focus on "carbon emissions," for instance, they're using metonymy to highlight a specific aspect of climate change, directing attention to a particular issue within the broader environmental context. Metonymy, while not something we may know much about, is regularly used in practice and often contributes to how various narratives are constructed around important issues.

Restorying

Through the process of co-creating worlds, we now set the stage for restorying personal or societal narratives. By shedding light on stories and narratives that perpetuate unsustainable practices, we're often more compelled to *rewrite* them with greater ecological awareness by storying the discourses that shape certain realities or perspectives. But as useful as storying can be for ecological awareness and action, it can also be employed by other forces to halt sustainability efforts.

In *Out of the Wreckage*, a book as relevant as ever right now, journalist George Monbiot argues that global politics needs a new, positive narrative rather than one driven by reaction and opposition. He identifies two predominant and oppositional political narratives from the twentieth century that still influence much of the world today: the social-democratic vision, which emerged in response to the Great Depression and then the Second World War, and the story of neoliberal capitalism, which emphasizes individualism, ownership, and economic market forces over shared societies and resources.[14]

The first story focuses on collective well-being and public good, attributing the cause of the Great Depression to greedy economic practices, such as tariffs, and a lack of regulation of business and financial markets. The second, neoliberalism, views the social system as an impediment to progress, advocating for individual freedom through reliance on market-driven forces and private interests. In some ways, these two stories can be seen as representing a choice between building a world for mutual benefit or creating one driven by private gain or wealth and power at the expense of collective, ecological well-being.

Both stories feature a heroic narrative with winners and losers, but one relies on fear and domination to attain our economic prosperity, while the other focuses on a more utopic vision: the social good leading to a better world. Regardless of historical effectiveness or accuracy, neither will continue to have relevance, leaving us without a unifying story for our future and contributing to our collective uncertainty.

Based on this explanation, Monbiot emphasizes the need for new social and political stories, as they are intertwined with personal narratives. Telling generative stories and learning how to tell them can influence people across the political spectrum. As Monbiot notes: "Those who tell the stories run the world." In that case, one might ask:

What's the ecological story for the twenty-first century?
And how can we collectively tell it?

These new stories emerge out of efforts to define who we are as people and societies. This process justifies particular actions or beliefs, and it can also reinforce larger master narratives that have unconsciously influenced societal norms for long periods. While understanding storying as a fundamental process to how humans have made sense of the world, it's also a dynamic process that can shift depending on the stories we expose ourselves to or convey to others.

One example demonstrating the trend of "those who tell the stories run the world" is how oil companies have a history of creating climate misinformation campaigns to sow doubt about the scientific consensus on global warming to protect their long-term financial interests. There's the case of Exxon-Mobil, which has been accused of funding climate denial organizations while internally acknowledging the reality of what climate scientists warn about. Investigations have revealed that ExxonMobil knew about these risks as early as the 1970s but chose to downplay these findings publicly and fund groups that spread misinformation to cast doubt on climate science.[15] These intentional disinformation campaigns have purposely created public confusion, stalled decisive action on climate change, and enabled oil companies to sustain their profits from fossil fuel extraction.

The use of narrative by corporations and politicians with private interests to sway public opinion is not a new phenomenon. One particularly misleading narrative is the idea that individual actions alone can solve large-scale environmental problems, ignoring collective or corporate accountability. As we show in the Relating and Witnessing chapters, there's a tension between individual versus relational responsibility. While individual efforts such as recycling or reducing personal waste are important, they're insufficient on their own to address systemic issues like climate change or biodiversity loss. Using stories as a way to blame others not only places unrealistic expectations on individuals, but it also diverts attention away from the need for systemic change, such as policy reform or corporate restructuring.

The master narratives around global environmental stories, such as the continuation of fossil fuels as the primary energy source or individual versus corporate responsibility, can be rewritten. Such a process of *restorying* creates opportunities to craft counternarratives to older dominant versions that no longer, if they ever did, support sustainable societies.

EXPERIMENTING WITH POSSIBILITIES

What if the dominant story we all shared wasn't about clinging to polluting fossil fuels but about embracing ecological sustainability and shared resources? What if the stories we tell about climate change and energy use focused on regeneration and interdependence, countering the misinformation campaigns of oil companies? By rewriting or *restorying* these narratives, we could empower communities to demand real change, not just in policy but in everyday choices. What if we could actively co-create this new story? What would that look like? What if each of us, through the stories we tell and listen to, could be part of shifting the master narratives that have for so long driven ecological destruction and social inequality?

Activating empathy

What are often called "climate stories" are increasingly popular ways to convey personal and social experiences that can inspire change. These narratives range from genres like "cli-fi" (like sci-fi only on climate fiction) and "solarpunk" (which imagines hopeful energy alternatives to dystopian futures) to personal accounts of extreme weather impacts. Often overlooked as too subjective, and therefore effective vehicles for change, these stories play a central role in building empathy and helping audiences relate to the urgency of climate or other ecologically related issues.

One reason for this is because stories have a profound impact on the chemistry of our brains. Neuroscientist and author Paul Zak has discovered that storytelling triggers responses in audiences that enhance their capacity for empathy and connection with others. Zak argues that the release of oxytocin when experiencing stories increases empathy, leading people to be more inclined to act altruistically.[16]

This phenomenon, also called *narrative empathy*, highlights how both consuming and sharing stories can expand our ability to understand and empathize with others. Whether we engage with narratives through literature, digital media, oral stories, or personal anecdotes, everyone can actively develop empathy and broaden understanding of the world.

By sharing our own stories, we contribute to the rich tapestry of social narratives, challenging dominant belief systems and paving the way for more inclusive and generative perspectives. This process can strengthen our resolve while also nurturing our capacity for empathy and creativity, which are essential when imagining different futures and restorying change.

We seek out and consume stories not only because they help us to imagine what is outside of our own experience but also because we derive enjoyment from them. After all, entertainment can be a motivating factor in social change. It happens all the time. One example is the animated film *Wall-E*. Frequently mentioned by the teachers we work with because their students respond well to it, as it vividly portrays a dystopian Earth that's polluted and abandoned, this film highlights the consequences of environmental neglect and unchecked consumerism.

> Engaging others' stories ultimately strengthens our ability to empathize, an increasingly necessary skill with fears and anxieties about the planet's future on the rise.

Let's keep building on this. Neuroscientist Marco Iacoboni, known for his research on mirror neurons, explains how observing and understanding other people through different mediums, like visual and digital culture, deeply impacts viewers. Mirror neurons are special cells in the brain that mimic the actions we see in others, essentially allowing us to *feel* their emotions by recreating them in our own brains.[17] This neurological process helps explain why storying is so important: it taps into our natural ability to empathize and connect with the experiences of others, even when they're portrayed on screen.

As we're seeing, storying becomes a convincing way to influence our emotions and attitudes, and how we perceive the world around us. With the research on mirror neurons and narrative empathy, there's now neurophysiological evidence for why we have historically been drawn to diverse forms of storytelling. Whether we're experiencing our own stories or those of others, our brains respond similarly, attuned to the emotions conveyed through virtual experiences.

———— ∘∘◯∘∘ ————

Another effective example of ecological storying is through conservation awareness campaigns that use motivating narratives to engage and educate the public about the importance of biodiversity conservation. Organizations such as National Geographic and the World Wildlife Fund (WWF) often use storying techniques to raise awareness about endangered species, habitat destruction, and other threats to biodiversity. By highlighting the beauty and significance of ecosystems and the unique species that inhabit them, storying can activate empathy and encourage people or groups to support conservation efforts, whether through donations or advocating for policy change through petitions.

Ecological and climate stories often feature charismatic species or emblematic habitats to evoke emotional connections and inspire action. A way to illustrate this is through "The Story of the Hummingbird." Famously recounted by Kenyan activist and 2004 Nobel Peace Prize laureate Wangari Maathai, it has also been retold in various global Indigenous traditions. We use the version inspired by the Indigenous Quechuan legend and complemented by Haida-manga illustrations in Michael Nicoll Yahgulanaas's account in *Flight of the Hummingbird: A Parable for the Environment.*[18] This version conveys how ecological storying combines word choice, metaphor, and metonymy with more-than-human actors, resulting in a form of narrative empathy.

Once, a large fire swept through the forest, forcing all the animals to flee to the river's edge, the safest place. They watched helplessly as the fire consumed their homes in the grove of spruce and pine trees. Bear, Wolf, Raccoon, Squirrel, and other animals expressed their sadness at losing their homes, sitting in despair and feeling powerless.

The only animal who appeared unfazed was Hummingbird, who declared, "I'm going to do something about the fire!" While the other animals sat by the river in safety, Hummingbird flew over, picked up a few drops of river water in their beak, and rushed back to drop them on the fire. Hummingbird repeated this process several times while the other animals just sat and observed.

Bear, the pragmatic leader, eventually said to Hummingbird, "Don't bother. It's no use. You're too small, your beak is too tiny, and these drops of water will not extinguish the fire." Undeterred, Hummingbird continued collecting drops of water and dripping them onto the fire. Despite Bear's discouragement, Hummingbird persisted in the mission. When Bear demanded they stop, and asked why even bother, Hummingbird responded, "Because I'm doing everything I can."

This brief story uses animal characters metaphorically, a storying technique called *personification*, to illustrate how each person, in their seemingly small ways, contributes to changing the world. It also emphasizes that despite perceived forms of personal and social apathy, we must nevertheless carry on and do the best we can in each moment, retelling stories that can transform collective narratives and change our futures.

Hummingbird is a metaphor for action and change, as well as courage and hope. The story teaches that all efforts matter, and every small action contributes to the energy and momentum of global collective healing. It also implies that the fire is but a symbol, a symptom of a larger systemic problem. Trees refer metonymically to ecosystems and biodiversity. Large movements begin with small grassroots gestures, like tiny drops of water on a blazing fire, to awaken collective support from others.

When faced with overwhelming challenges, it's common to become paralyzed by fear and inaction. By sharing and amplifying ecological stories like "The Story of the Hummingbird," people and communities can find hope. Inspired and motivated to tackle risk, the story contributes to collective efforts to build a more sustainable and resilient future.

PRACTICING

On your own or in groups, explore and share some ecological stories that resonate with you, whether personal accounts of climate impacts or narratives from literature, films, or folklore like cli-fi or solarpunk. Start by reading or watching a story, such as "The Story of the Hummingbird" or *Wall-E*, and then reflect on its emotional impact. Then, share other ecologically related stories, either real or fictional, and dialogue about how these stories might connect to broader environmental issues like consumerism and waste, biodiversity and species extinction, or weather events like fires and floods. Also, consider exploring how stories, through narrative empathy, can build a collective awareness and action.

Reading the environment

Storying our experiences of ecology is just one of many examples that can help to inspire and illuminate the dark and anxious responses we have to changing the conditions and the actions we take in the world. With the dominance of scientific language and political debates that can often exclude many groups of people, we become passive recipients of news rather than agents of change.

Traditional knowledge passed down through stories has given way to fact or statistic that typically excludes the narrative context from which those facts emerge. For example, English has started to lose the vernacular sayings that were once a common means of communicating knowledge about the environment. Variations of the folkloric expression "red sky at night, sailor's delight; red sky at dawn, sailor be warned" were used extensively to convey basic knowledge about how to *read* environmental phenomena. Other similar expressions included "if birds fly low expect a storm to blow" or "when sounds travel far and wide, a stormy day will soon betide," which now seem quaint in their language use.

Some of these phrases still hold truth but uncertainty and the changing narratives of climate science and weather prediction have made such folkloric knowledge almost obsolete. The ability of stories to reengage traditional folkloric knowledge, based on lived experience, can help us share our knowledge and repair our ability to read and respond to the environment.

Reading the environment so that we can story the interconnected phenomena that mark significant changes in our physical realities has never been more important. Stories communicate complex topics in a relatable manner. As such, storying can serve as a practice of advocacy and social change, mobilizing communities to address environmental issues through collective awareness. By sharing stories of successful environmental initiatives or highlighting the resilience of ecosystems, we can inspire hope and encourage people to engage in efforts to protect and restore the natural world.

REFLECTING

- Reflect on a story that has influenced your perspective on climate change or any kind of environmental issues. Was it uplifting? Scary? Nostalgic? How did it shape your beliefs or inspire any actions, if at all?
- Can you think of any metaphors or metonyms used to describe the environment – such as "war on climate change" or "saving the bees" – and how they might shape the way society might engage with ecological issues?
- In what ways do you think the use of storying in environmental campaigns, such as those by National Geographic or WWF, might influence emotional connections to ecological issues and motivate people to gain awareness or take action?

CHAPTER NINE

Witnessing

An adaptable education

As educators, we've spent years exploring how learning, and unlearning, can reshape our relationship with the world and deepen ecological consciousness. And yet, we're still figuring it out. As made clear in previous chapters, this work involves navigating language and meaning in ways that support a more sustainable planet. That challenge feels even more urgent now, especially in an era called the *polycrisis*: overlapping and interconnected crises that push social, ecological, and political systems beyond their ability to recover.

So we return, again and again, to a guiding question: how do we encourage ecological awareness and responsibility in times of such deep uncertainty? We ask this not just as teachers but also as learners ourselves.

Many of our students, rightfully so, would rather be out in the world *doing something* about these crises. And yet there are ways to educate both within and outside of formal schooling structures that allow for deeper understanding that can assist with systemic changes.[1]

Building on other concepts introduced in this book, such as becoming, listening, and composting, this chapter's focus on *witnessing* can be seen as a vital approach for lifelong learners and one that might be the essential piece in finding greater social unity through a shared sense of compassion and empathy. It might seem like a long shot in today's world, but when we change how we talk and act, an ecological shift isn't only possible, but it's also already in motion.

As both a concept and practice, witnessing brings about a language of awareness that goes deeper than seeing or hearing. It's all encompassing, by which we mean embodied and visceral as much as visual and concrete. This chapter engages with testimonials of *witnessing*, scientific practices of *observing*, and sociopolitical acts of *testifying*. Witnessing bears with it the responsibility to expose uncomfortable truths, even when one might feel endangered by doing so.

Regardless of whether what we witness is joyful or sad, testifying to it is an act for future generations. It's rooted in the belief that we are capable of doing better, acknowledging the past and present as they are while speaking up for an uncertain future.

Witnessing is also about place-based knowledge. It removes the layers of mediation and abstraction from experience and puts us in the present moment. Witnessing means letting go of our learned skepticism so we can fully absorb and respond to the lessons unfolding before us. It involves adapting. It teaches us to be present, as in responsive and responsible, and to keep an open mind to prevent filtering out details that don't conform to preexisting dominant viewpoints.

There's an important lesson to realize about adaptability in learning that applies to both language and ecological systems. *Morphology* is the term linguists use for the study of how words can change and evolve given a particular context of use. Language is highly adaptable, and we're witness to it constantly. This is especially true for common nouns and verbs in English that can change from being a noun to a verb or to an adjective or adverb. For example, we can *form a sentence that has good form*. Form, in this instance, is both an action (verb) and a thing (noun).

Fun trivia: the longest grammatically possible sentence using only one word in every position of the sentence is the word *buffalo* written eight times over: *Buffalo buffalo buffalo buffalo buffalo buffalo buffalo buffalo.* The first word is a proper noun, the city of Buffalo, New York; the second is a non-count noun of animal; the third is a verb to move something roughly, and so on. *Buffalo* is a very versatile word, while also having a deep ecological and cultural history in North America.

The point of this little experiment is that language serves as a dynamic and adaptive system. Language adjusts to an ever-changing context of use, and it provides for the possibility of infinite variety in expression. It can serve to inform, and it can serve to inspire stimulating the imagination. It can model what we strive for as we seek for long-lasting change while sustaining a functional society.

This chapter builds on the fluidity discussed throughout the book, focusing on how witnessing unfolds through language practices like testimonials, which relate to storying our experiences. It highlights our responses to both everyday crises and high-profile disasters – what we refer to as *commonplace* and *celebrity* catastrophes.

Shifting our thought pathways

Throughout this book, we've emphasized actionable "verbing" concepts, viewing them both symbolically and functionally as ongoing processes that require continual adaptation to changing circumstances. How we take up the adaptability of language, its morphologies, and integrate this in broader forms of education for various types of learners can have far-reaching consequences. In other words, this isn't just language play for the sake of it. It's deeply rooted in everyone's lives.

We highlight these morphological principles of language because, returning full circle to the opening chapter Discoursing, there are what Bruno Latour called matters of concern and matters of fact.[2] Communication based on *matters of fact* can easily be turned against progressive causes, particularly through misinformation or misappropriation. In fact, the use and misuse of facts have greatly hindered environmental movements, which are more appropriately considered *matters of concern*. Consider, again, the well-trodden word *climate*. It began as a matter of fact but has increasingly become a matter of concern, as the facts alone could no longer account for its growing social and emotional impact.

By emphasizing matters of concern, which don't require proof that can be manipulated but emphasize values and beliefs that can affect worldviews, we bring to bear another form of truth called *witnessing*. In witnessing something, there's already a language event occurring because it comes with the reciprocal exchange of giving and receiving testimony. The matter of concern is the subjective event or experience being witnessed. Sharing that experience is often done through contextual narratives, or *storying*, as we explored in the previous chapter.

As a genre that's about unmediated encounters with the real world, testimony focuses on truth telling, sometimes under oath, sometimes in front of other witnesses. It can also be storied and embellished for the sake of entertainment. Witnessing something important implies an obligation to speak and act, to *spread the news*.

Although language is highly adaptable to the circumstances of use, thinking isn't necessarily so fluid. Throughout our learning lives, we develop what are called *mental schemas* that help us store and organize memories of our experiences and knowledge, allowing us to recall and logically arrange them as we try to make sense of the world.

Schemas, as developed in literacy education, refers to the cognitive frameworks through which we integrate and comprehend information from reading, analogous to assembling pieces of an infinite jigsaw puzzle.

Schemas, or patterns of thinking with language, can also become rigid and unflexible; they become ingrained and difficult to change regardless of the context. Thinking becomes habitual. When neurons in the brain develop specific habitual pathways for thought, the nerves grow to accommodate those patterns, much like ruts forming on a dirt road where vehicles frequently pass. In other words, it's easier to drive down the usual path of our mental schemas than to blaze a new trail of thinking. Taking it a step further, it's often easier to stick with what we know than to change how we think or act.

An *ecological schema* could be the mental framework people use to understand and categorize ecosystems. When someone thinks about a forest, their schema might include trees, wildlife, plants, streams, and the interdependent relationships between these elements. They also might think about healing experiences of peace and tranquility. This schema helps them understand how a healthy forest functions and their relationship to it. Over time, schemas could be shaped by cultural narratives, education, lived experiences, and even media representations. So a forest might also evoke memories of childhood hikes, stories of deforestation, or spiritual connections to land. An ecological schema isn't static, though. It evolves as people encounter new information or experiences, and it can either support or limit how we act in response to ecological realities. Ultimately, schemas are how our thinking represents the interconnectedness (and interdependencies) of complex ideas and processes.

———— ooOoo ————

Schemas can be useful in many ways. They're a basic function of intelligence and making sense of the world. When information is encountered that doesn't fit the schemas one has in their head, if it's an anomalous situation or an irregularity to the norm, it produces what's known as *cognitive dissonance*. This condition is where psychological discomfort or tension arises when a person encounters information that substantially contradicts their beliefs.

Note how the metaphoric language here changes from schema as a visual concept to dissonance as a sonic concept. Dissonance is an internal state. It arises when a person holds conflicting values, when their actions don't align with what they believe to be true or right. It forces a process, large or small, of adaptation and an expansion of original schemas so that they no longer contradict or take on meaning in their contradiction. That's because what's at stake here is understanding. One can know two contradictory theories, but to understand them, one needs to make them fit in a broader schema.

In a sense, one needs to reorient oneself in relation to the ideas they contain. It can be so dissonant as to call into question the entire preexisting schema and that implies a lot of unsettling work, undoing and re-creating new neural pathways for that thought process to follow. Learning about the deep history of colonialism in national parks might be so dissonant, for instance, that it forces someone to rethink their long-held belief that these places are purely natural and apolitical.

Here we have the crux of the issue for environmental causes and communication today. We're headlong into what the literary critic Fredric Jameson called *late capitalism*.[3] Not *late* as in *it's over or deceased* but *late* in the sense that it has grown into *a fixed set of values*. In this case, the schema provides a rationale, an explanation for a particular economic system to prevail as an entrenched set of social norms.

Faith in the capitalist way of life is hard to unsettle because of its schematic rooting, one that's often contested with a false opposition of communism. Ecology is the true opposite of the dominant capital resource economy based in ownership and control. *Capital* refers to resources or assets that generate wealth or income, but it also represents the extracted and abstracted value from something. Ecology is quite different as the situated and interconnected values of the whole. This is amply demonstrated in language.

The language of capital is about competitively gaining property and ownership. It's equally about exclusion and predation. Market capitalism, epitomized by global stock markets, removes all context and physical consequences from transactions. Capitalism is an abstraction that has dire consequences because the players in this system often don't have to live with the consequences of the actions that have created their wealth. Those consequences get passed on to localized, situated lives, often those populations and species with no direct benefit of the wealth being generated.

This dominant model of global societies is increasingly causing cognitive dissonance. It's like the rules of the game of life in contemporary society are changing, and the so-called *winners* are, unsurprisingly, not going to let that happen under their watch. Sound familiar in the current global context?

The problem is that people are now witnessing environmental change through extreme climate events and ecological crises, which creates dissonance with their beliefs in capitalism's efficacy, the naturalness of the social order, and systems of human dominance. An example of this would be someone who has relied on the language and actions of technological progress might feel deep discomfort when they see repeated wildfires, floods, or species loss that technology alone can't fix.

There's nothing like witnessing to either *unsettle* routine thinking or instill motivation. When we see, hear, smell, and taste firsthand, we can no longer sustain denial. Witness narratives are highly valued source of information in both environmental and legal discourses, but they're also a constant feature of commercial ones as well. Referring back to the economic language, this is akin to reading consumer product reviews to get a firsthand account of the product in question.

Witness testimonies bring us closer to embodied experiences through language, cutting through the layers of mediation and interpretation that typically accompany the spread of information.

But witnessing also implies a duty to tell others. The witness is called upon to speak the truth because they were in proximity to a given event. They were a part of the ecosystem of that event when it happened. Witnessing is therefore a spoken event because of its form of testimony. Witnessing gains relevance in the telling.

Testimonials, like user reviews, play a significant environmental role. Businesses often downplay environmental harm and health impacts because they affect profit. Using general language about issues like logging or mining often lacks emotional sway and doesn't connect with those uninvolved. The testimonial, as a personal matter of concern, not only bridges any potential divides through a form of language that's as ancient as stories but also offers a direct rather than imagined connection to events.

As a genre of expression, testimony provides a guarantee of authenticity, though that guarantee is often abused in corporate communications and marketing. Even if we weren't there to witness something firsthand, hearing the story from a survivor draws us closer to the experience through their language.

Rachel Carson's book published in 1962, *Silent Spring*, is an effective example of the impact of testimonials. The book, which exposed the cata-strophic effects of chemical pesticides on the living environment, was inspired by a letter from Carson's friend, Olga Owens Huckins. In 1958, Huckins wrote about the dead birds she saw on her property in Massachusetts after DDT (dichloro-diphenyl-trichlo-roethane) was sprayed nearby.[4] This letter sparked Carson, who was trained as a marine biologist, to conduct research and then write a book that helped ignite the environmental movement in North Amer-ica. Witnessing, in this case, cuts through disinfor-mation, revealing the truth hidden by industry cover-ups and placing importance on subjective meaning involving *matters of concern* in our awareness and understanding.

The act of witnessing does carry with it a cost. As we've mentioned elsewhere, there's a delicate balance in ecologi-cal discourses between using language to inspire empathy and drive action across diverse groups and using language that inadvertently discourages people, hindering their abil-ity to change. This is a very fine line for people who consume media daily because it's digging deeper ruts of schemas that revel in catastrophe.

Catastrophe looms not in the next century, decade, year, month, week, or day. It's quite literally here now, on our doorstep, in our house and beds. We're witnessing these *commonplace* and *celebrity catastrophes* everywhere, all the time. These catastrophes convey certain meanings and impacts that are shaped by and rooted in language.

PRACTICING

In a group or on your own, reflect on a local environmental change you've witnessed or experienced, such as the disappearance of a public park, the loss of an animal close to you, or the transformation of a natural landscape, like a local wetland or forest. Write or dialogue about your memories of what was there before and what has been lost. Also, consider what role you could play in witnessing and then advocating for change. What would your testimonial look like? Consider sharing your reflections, talking about the responsibility of bearing witness and the impact of these losses on both the environment and community.

Commonplace catastrophe

The term *catastrophe* derives from ancient Greek terms meaning down (*kata*) and turning (*strophe*). It comes from a culture that explored the notion of catastrophic events in their tragic literature at great length.

For the ancient Greeks, this term extended beyond a literary form of theater, namely, Tragedy, to a graduated place in the grand narrative of human civilization. The literal Greek meaning of *strophe* was to turn up and down, what has been characterized as a hero suffering an ultimate reversal of fate or circumstance. It implies a sudden overturning, the result of which is death of the person, dream, ideal, love, planet, or promise of a good and just future. It marks a finality, an end of all hope for a different or better outcome.

This change in fortune, as elucidated by the epic poets, was of metaphysical proportions, and included the fortunes of the gods and spirits that filled tragic theater. Tragedy is also a commonplace term. While we use the word tragedy to mean losing something meaningful or special, such as a death of someone close like a family member or friend, the meaning of *catastrophe* retains some of its epic, ultimate reversal of fortune connotations.

Catastrophe is about as bad as it gets because it marks the end. No more chances. Finito! It also marks transition into something else, a vacuum from decimation waiting to be refilled. As Aristotle observed, nature abhors a vacuum. The same could be said for political power shifts in society.

This word *catastrophe* is closely related to another word of Greek origin, *cataclysm*, which derives its value from the notion of a downpouring or washing away. It's a meaning that's closely related to the flood story of the Old Testament, which was a retelling of an older Mesopotamian epic of a time when the known world was submerged in water.

Cataclysm still has uses today in geology to mean an event of great magnitude, like a major earthquake that causes uncalculatable damage. With rising sea levels, caused by escalating temperatures, the balance of ecological systems that characterized life through the human-created Anthropocene has passed, and we're in a time of terrestrial turmoil, a period of cataclysms, the experience of which portends catastrophe on the human scale.

But the story doesn't end there. Christianity then came along and took some of the sting out of catastrophe. The New Testament of the Bible is an epic tragedy that twisted the plotline in a favorable way. Just continue believing in the hero of the story and you'll get a second chance, an eternal life.

Take John 3:16, for example: "For God so loved the world, that he gave his only Son, that whoever believes in him should not perish but have eternal life." Or read Peter 4:8: "Above all, keep loving one another earnestly, since love covers a multitude of sins." John 4:8 affirms "God is love," and Corinthians 13:7 claims love "endures all things."

It's easy enough, from this perspective, to see why this spin on the narratives of old was so popular. Even if catastrophe occurs, you can just go on believing, and it will all be okay. You'll escape with the promise of eternal life even if the planet combusts in flames.

Catholicism, which could be seen as the prototype corporation, went one step further. It added an economic system of governance to faith, using the belief or faith in God like an accounting service of divine love. In essence, it developed the first and most profound personal futures market, where one could speculate on eternity if one had the wealth and resources. By giving money to the church, one was promised eternal life after the cataclysm that marked the end of days for humanity.

Fast-forward about 2,000 years and it's easy to see why politicians and leaders in the West give so much lip service to Christian faith. It reassures people that no matter what, if they stay true to that storyline, even the worst events will be managed by faith alone. It encourages many to follow leaders with *the right story* with unwavering loyalty.

This brief history is part of what makes the current planetary situation so challenging. The cognitive dissonance is becoming thunderous: the subversive faith in eternal reward, in miraculous renewal, which contained the depressing implications of catastrophe like casks for radioactive materials, has cracked, and the disguised meanings are leaking into collective consciousness. We have come face-to-face with destiny, witnessing or hearing testimonies of catastrophes that happen on a continual basis around the globe.

But here's the thing: it's not the "wicked" who are being destroyed. More often it's the hordes of innocent people, and even further the more-than-human life-forms. People are witness to catastrophic change, regardless of where they live on planet Earth. The catastrophe becomes a tragedy when we can't do anything about it. *Catastrophic change* means something is coming to an irreversible end. It's not like tearing down an old building and putting a new one in its place but more like rebuilding a city after a war. This is a time when the belief in renewal itself must suddenly come under intense scrutiny.

Witnessing commonplace catastrophe has become unavoidable, and the older people are, the more tragic environmental change they might have witnessed. Indigenous ways of passing on understanding and tradition place elders in precious and respected roles as knowledge keepers. Similar practices occur in other ancient global cultures.

This is in stark contrast to Western practices. Very few, typically wealthy seniors, are afforded respect. The tragedy of this is that some of the most remarkable environmental heroes are senior-elders, from all walks of life, motivated by witnessing and communicating commonplace catastrophe daily.

Commonplace means a shared experience of the mundane (*mundos* = of the earth) kind of catastrophe, as the end of something good. This may include the disappearance of flora and fauna, not for a season but forever; the destruction of a favorite environment for a new suburban subdivision; or the vanishing of spring frog song when the bogs and marshes are paved over for box store parking lots.

These kinds of catastrophes go unnoticed unless someone bears witness to what was there before. These catastrophes occur daily and accumulate. Elders around the world hold the key to change because they can give testimony to what has been lost over a longer duration of time. And holding space for that loss is part of futuring for ways that support our planet's health.

Celebrity catastrophe

The celebrity catastrophe is the type of global event that gets a name and mainstream media spin. These catastrophes attract profiteers to take advantage of recovery efforts giving rise to *disaster capitalism*, which Naomi Klein expertly describes in her book *Shock Doctrine*.[5]

Disaster capitalism harnesses a kind of missionary zeal. In a state of emergency, governments of the affected area will be in no position to negotiate terms for recovery efforts. They must conform and convert and ultimately pay to receive help.

The *celebrity catastrophe* is a catastrophic event that's celebrated, even though the celebration is full of mourning, and thus, it differs in value, if not in kind, from the *commonplace catastrophe*. One has more value because of its cultural and financial gain, or the ability to turn this catastrophe into a marketable and wealth generating opportunity. It's the kind of news that grabs attention in an economy built on profiting from it.

While the commonplace catastrophe can go unnoticed, or is restricted to regional reporting and concern, a catastrophe of more global proportions is difficult to ignore. These catastrophes, such as the West African Ebola outbreak, Hurricane Katrina or Harvey, the Fukushima nuclear powerplant failure, Chernobyl, COVID-19, and so on, have typically been measured in terms of loss of human life and property. This gives rise to increased media attention.

What makes these events different from commonplace catastrophes is that they happen suddenly and notably, and they affect humans directly. It doesn't matter whether they're so-called natural or human-caused disasters. They're still catastrophes in the epic genre.

One of the main challenges today in responding to catastrophes, understanding their deeper meaning and adjusting human behavior to prevent them, is that witnessing these events has largely become a form of sensationalized entertainment for viewers. Witnessing as entertainment doesn't give over the same level of personal commitment as witnessing does when it causes a relational bond, when one *bears witness*. Testifying carries with it a legal and psychic burden, so we tend to shy away from it. Commonplace catastrophes aren't ones we typically want to bear witness to because they are, to put it bluntly, depressing and not entertaining in the slightest.

A United Nations report warned that over one million known species face extinction, with rates now hundreds of times higher than in the past million years. One species vanishes every two to three weeks.[6] These are the commonplace catastrophes that we're blithely unaware of. But they're total catastrophes just the same. And this doesn't even include all those *persons* that otherwise make up the global community, such as the disappearing old-growth forests. Each forest that is cut down and turned into an impoverished tree plantation is as much a catastrophe as the loss of a species. For there will be no more 1,000-year-old trees, from here on because that process can't be sped up with any form of technology.

Now we spend time bearing witness to death, involving loss and disappearance, instead of bearing witness to the magical and spiritual entities and properties that inhabited wilderness places, that kind the Lakota storyteller Black Elk speaks about. Interestingly, Black Elk attempted to help some Harvard-educated researchers grapple with the notion that spirits are real. Before a human goes messing with a place, they had better negotiate with the spirit of that place first. For those Ivy League thinkers, brought to witness the living persons of stone, starlight, tree, or fur, it was in the practice of witnessing that they encountered profound life-changing, schema-destroying dissonance with all that they had known before. Indigenous knowledge keeper and scientist Leroy Little Bear did the same for the reporter who struggled to see the spirit energy in southern Alberta's boulders and the communication practices that arose from them.[7]

Why does this even matter? Is this just another form of passivity? Witnessing is vital to effective change because it places an inescapable burden of responsibility on the person or people witnessing. Though they might never act on it, being a witness means being bound in an ineffable relationship – one that's entangled in the web of discourse as the bearer of testimony. Whether as victim, aggressor, or bystander, one carries the burden of catastrophe and the knowledge of the downturn or a profound and irreversible end from which there is no return.

This is why economic, social, and even the overlooked spiritual system of capitalism is so effective. The global economy and multinational corporations enable those who profit most from destruction and catastrophe to remain detached from its consequences. Or, to put differently, they never bear witness to it.

To play the stock market doesn't require one to take responsibility or witness the everyday realities of the workers. The lower down the chain of power and command, the more one must bear witness to the darkest of corporate endeavors, to the destruction of vital forests, rivers, marshes, mountains, and all that attend them. In short, to do the *dirty work*. It's dirty because it comes with a stain that can't be removed: the stain and strain of witnessing without providing testimony.

To prevent witness testimony, many people are forced to sign nondisclosure agreements, silencing their ability to speak or communicate. But they must then live with what they have witnessed in silence, and to do so is to take on responsibility for the ongoing consequences into the future, *into the silent spring*.

We might consider placing witnessing in a special category of discourse ecology. Witnessing, especially around environmental issues, is denigrated in a most particular way because it breeds necessary action, which can manifest socially as activism, among other ways. This gets in the way of economics based on profit and ownership, which try to turn everything, even the whole Earth, into an abstract system of value called money. They don't see that the value is already there before the resources are taken and converted into money.

We're foreclosing our own future. Without bearing witness to astonishing experiences as well as experiences of loss or extinction, there will be no solidarity with environment.

All of this returns us to celebrity catastrophe and its ability to capture our attention because of the forms of digital, social, and mainstream media, and resources it brings with it, while also making it a form of entertainment. And we can see from popular movies and TV series how compelled people are by the return of the epic tragedy, with all the apocalypse-themed narratives that proliferate on popular streaming sites.

One reason these narratives are so popular is that, as a society, we feel the need to witness and confront the changes happening around us. Only now, the epic proportions of these catastrophic events are not just imagined, they're very real in both local and global contexts.

It was said that the Roman emperor Nero sat on the walls of Rome playing his violin while watching the city burn, which was the center of Western civilization at the time. In other words, he saw catastrophe as a kind of entertainment. Are we doing the same with the planet? The antidote, it would seem, is to provide equal space for witnessing testimonies so that we can allow this information to motivate real-world change.

EXPERIMENTING WITH POSSIBILITIES

Consider a world where every ecological catastrophe was witnessed by every person in a community, and their testimonies had the power to transform society's relationship with their environments. What if witnessing became the central act of education, moving beyond abstract theories to lived or observed experiences? This thought experiment invites you to explore how the act of witnessing could challenge rigid schemas of thought, such as those shaped by the economic beliefs rooted in ownership with unequal wealth distribution, and create a new language that bridges the gap between knowledge and action. How could stories of environmental loss reshape societal values, promoting an ecological mindset and responsibility? Could this *witnessing* challenge entrenched economic systems and foster a collective shift toward recognizing our interconnectedness and the urgency for change?

Restorative listening

Witnessing puts us wholly in the present and reintegrates us, willingly or unwillingly, with the local environment. It calls upon us to adjust our thinking when what we have witnessed produces a cognitive dissonance with what we believe. When many people witness something unlawful and no one is willing to speak up and act, we call this *bystander effect*. Bystanders displace the burden of care onto others who are also witnesses. If no one responds, if no one bears witness, change is much less likely to happen.

Bearing witness is like shifting our role in language syntax, moving from a noun to a verb, testifying to what has happened and becoming the change we need to witness. When enough people bear witness to catastrophic events, change is already happening.

That's why witness testimonies were the core of the Truth and Reconciliation Commissions in Canada and South Africa. They began their work through forms of restorative justice, hearing the firsthand stories of survivors to understand the impact of colonial systems on lands and people. A way of resolving conflicts or addressing harm, *restorative justice* emphasizes repairing relationships and addressing the needs of all parties involved, rather than solely focusing on a zero-sum game of punishment mediated by blame and shame.

To reconcile an irreconcilable experience of witnessing atrocities, one must speak about them to others. Such expressions of speaking may take many forms, including art, sound, and bodily movement, extending beyond conventional verbal communication. Others must listen to those narratives so that they lift and carry some of the burden that's borne by the witnesses. This process heals and repairs, relating by creating meaning through the many layers of communication.

Witnessing the real world, with all its damage and the cascading collapse of ecological systems, doesn't have to be all about tragedy and inevitable catastrophe. Witnessing can inspire change by forging necessary adaptation and recognizing the irreplaceable beauty found in the natural world.

When we witness something glorious, full of awe and wonder, it inspires us to act, turning witnessing into a powerful motivator. Because of this, witnessing also speaks to the inexpressible joy of profound encounters with the diverse environments around the planet. It points to the first time we witness a butterfly emerge from its chrysalis, hear a meadowlark, see a squadron of manta rays, or experience a stunning desert sunset.

In this sense, witnessing is renewal and reawakening ourselves to the awe and wonder of youth. Witnessing need not be about catastrophe to be an effective way to change and inspire new hope in how we can act and talk about ecological collapse. We can speak of all we witness in the natural world, recognizing the profound loss it would be for future generations never to experience its wonder. We might try to witness with the eyes of a child and testify with the wisdom of elders who know what it is to lose something wondrous and precious forever.

REFLECTING

- In what ways can witnessing, through personal stories or testimonials, help shift our understanding of ecological collapse and motivate us to take responsibility for change?
- How does the difference between *commonplace* and *celebrity* catastrophes affect the way we witness and respond to environmental change?
- How might we make space in society for deep listening and for collective witnessing? Might we all give testimony to what we most treasure about the Earth?

CHAPTER TEN

Futuring

The imminent now

Predictions of the future loom large for every one of us with finite time on this planet. What do our futures hold? How much control do we have over these potential futures? Is it even possible, at this point, to imagine and build more sustainable futures? These are the questions that keep many people up at night. They link to so many of our hopes, fears, anxieties, and aspirations for the planet.

Thinking about the future tends to generate reactions of anxiety and fear for many. The future represents the unknown, not yet known, and, quite possibly, never known. Let's not forget that it can also provide infinite possibility, where the curiosity of how life continues to emerge in mysterious ways. The tension between anxiety of the unknown, ultimately generating what has been pathologized as *climate anxiety*, and the hope found in imaginative possibility is key to living in the moment while also building a better future.

As we've navigated these corridors of paradox, the specter of the future has been present throughout our conversations. This final chapter serves as both a continuation and a conclusion, exploring *futuring* as an ecological reality in our lives. Building on all the chapters throughout the book, futuring is about imagining and anticipating possible futures based on our collective creativity and experiences, showing how the connection between communication and existence can be expansive.

But the future is now, in this moment, the timeless present. It's also in the past and the future, experiencing time as fluid and spatial as opposed to linear. Because the future is always unfolding in the present, language helps capture what the mind struggles to grasp.

We might consider the gradual shift from "global warming" to "climate crisis" and from "mass extinction" to "ecological collapse" as the ongoing examples. Changing the words can alter how we see an ecological future and our behaviors to it as a future event also happening now. In this way, the concept of *futuring* captures both perceptions of time and our relationship with the planet.

Futuring is, at its core, a language of practice and of possibility. But it's not just words, nor is it limited to actions. So how does language contribute to creating possible worlds or worldviews, particularly ones that align with ecological practices? This is the grand futuring question, one that's as uncertain as any future, but it's nevertheless an underlying focus of this book: a practice of becoming ecological through the many discourses we adopt and adapt.

Time's trick

That slippery thing called time that binds us all is a puzzle that many of us try to make sense of through language and thought. In reality, futuring is mostly about psychological *mind time* because the biological *material time* elements of our existence don't rely much on the future or the past. If we pay attention, we can see how our bodies live in the moment, each day, eating each meal, regulating temperature, and needing a toilet. Bodies function in autonomous ways without the need for our attention, our thinking mind.

Enter stage left fear and anxiety, which emerge out of this morass of abstract thinking in psychological time that occupies much of our mental headspace. This is true of climate doom as well as any other threats to our existence, whether that's nuclear annihilation or a car accident.

The relationship between psychological time and Earth's ecological realities is multifaceted, but understanding it could help manage future-related anxiety through our language and actions. *Psychological time* refers to the subjective experience of time influenced by perceptions.

Something like polar ice cap melting, in contrast, represents a material reality and a global environmental challenge with far-reaching implications for ecosystems and other life on this planet including future generations. While material time is real, psychological time is just a story in our mind.

The concept of psychological time is drawn from a series of dialogues – later published as *The Ending of Time: Where Philosophy and Physics Meet* – between the Indian philosopher Jiddu Krishnamurti and American physicist David Bohm, in which time was a central theme. The complementary nature of material and mental experience mirrored the synergies between a philosopher and physicist. Material or biological time consists of the body involving physical life and death. Regardless of what we think or do, our bodies move forward through time in a somewhat linear way.

Abstract or psychological thinking functions outside of linear time. One might even say that time ceases to exist, even though it doesn't for most of us and often dominates our existence. As Krishnamurti observed, "the idea of tomorrow doesn't exist psychologically." In its most basic sense, "thought is time," but it's constructed time as a process of thinking that's related to one's perception of reality.[1] This is why people often experience time as having different speeds. When we're in love, time moves quickly. But if we're at the dentist, it drags on.

Shifting the concepts of "I" or "self" to "relating" also emerges out of psychological time, making language a key element of this perspective. Time loses its grip once we realize it's mostly shaped by how we perceive reality through our thoughts, particularly by the language we use to try to make sense of it.

This matters for language ecologies because of how we think or talk about time shapes our sensations and reactions to circumstances like air pollution or heat waves. When changes happen slowly, it can make us feel apathetic, in denial, or powerless, especially if the future seems uncertain. These feelings can hold us back from taking action because they lower our motivation.

If climate change is talked about as a distant concern sometime in the future, people might not feel like it's urgent because it's actually happening simultaneously right now and in the future. But if we use language that emphasizes our direct connection to ecological issues, something akin to *wildfire smoke is making the air unbreathable*, then it can inspire people to act in real time.

Such extreme weather events like forest fires or hurricanes can serve as *temporal landmarks*, marking significant moments that disrupt the perception of time and prompt people to reevaluate priorities and behaviors. Effective communication strategies that frame climate change in terms of immediate risks and tangible impacts like melting icecaps or drought can also help bridge the gap between psychological time and the long-term effects of what these changes imply.

The 2021 heat dome in British Columbia, Canada, with record-breaking temperatures and wildfires, turned climate change from a distant threat into an immediate crisis. The town of Lytton hit 49.6 degrees Celsius (121.3 degrees Fahrenheit) before being destroyed by wildfire. Wildfires burned 8,500 km². In this extreme weather event, caused by a high-pressure system trapping hot air over the region, over 600 people died, mostly the elderly or the isolated. Hospitals were stretched while infrastructure failed. For many, it shattered the illusion of stability and safety, revealing how quickly everyday life can be disrupted by extreme weather. As an event that we ourselves lived through, we can attest to how this temporal landmark spurred immediate public discourse on emergency preparedness and policy action.

One aspect of this relationship lies in the language of urgency and immediacy. Climate change often unfolds over long time scales, making it difficult for people to perceive its immediacy and prioritize action in the present. People tend to prioritize immediate rewards over future ones, a bias called *temporal discounting*, making it harder to push for practices or policies that tackle long-term changes. For example, even though transitioning to renewable energy is both environmentally and economically beneficial in the long run, many governments and industries still invest in fossil fuels for short-term profits and predictability.

Our day-to-day existence is characterized by time speeding up. This is becoming universal. With digital devices aligned by satellites to atomic clocks set so that the whole world ticks away in synchronized measures, we start to treat time as a thing, as a noun. It becomes the universal object called time, a capitalized Object just like Time. Capitalized, in this sense, has multiple meanings. We quite literally waste, spend, and buy time, as reflected in the phrase "on our *own* time," which suggests psychological ownership. Time becomes something we compartmentalize and commoditize.

This contrasts with relational time, which unfolds through dialogue and exchange. Dialogical time is durational, which encourages listening and witnessing, and rushing it risks losing meaning altogether. In contrast, we process visual information rapidly, making instant judgments about whether it's worth our time and attention. It's fed to us at dizzying speeds and is outdated almost as soon as it arrives.

Spoken expression and listening are locked into relational time. Speech is a matter of *timing*, paramount not only for public speaking and comedy but also for intimacy in the moments of communicating. Our sense of time is affected by the speed of our communications. As our devices get faster and discourse happens more rapidly, we have the sensation that time itself is speeding up and that the future is always here, already shaped and unchangeable.

The distant future gets harder to see with the present so present and the new so new in communication systems. As many of us experience, technological speed and volume of communications changes psychological time. That could be affecting judgments at all levels of personal, social, political, legal, industrial, and, of course, ecological. Making good judgments takes time so we can consider many different viewpoints. The apparent closeness of the immediate future obscures our envisioning and storying of more distant futures.

How might we envision long-term futures shaped out-side the pressures of speed and the constant race against time that currently defines our realities?

Throughout this book, we've pointed to some fundamental aspects of written and spoken language, visual information, and vision-oriented thinking that are having some very unfortunate consequences for ecological systems. This moment requires deep *listening*, *storying* the present, *relating* to each other by *witnessing* and giving testimony to lived experiences, *ingesting* a diet of richly varied information, *composting* cultural excess and recomposing it, and *unsettling* our social habits and expectations to set forth on a sustainable path. These practices of our ongoing ecological *becoming* through language can help to stabilize the psychological effect of rapid time.

Understanding how psychological time influences attitudes and behaviors toward everything related to the collapse of living ecosystems across the planet is essential for developing effective mitigation and adaptation strategies that resonate with diverse audiences and promote collective action. In environmental discourse, there's always a lot of futuring going on, but much of it is seen through a dark lens of immanent catastrophe: we seem to be lacking depth of experience and positive possibility.

The change we seek is material and concrete, as well as psychological and perceptual. Although language systems aren't material systems like oceans acidifying or crop yields deteriorating, they can hold powerful sway in how humans respond to polycrises by changing how we talk and act. This is the ultimate contradiction of our *time* – saving our possible futures may depend on the language we use to imagine them becoming.

EXPERIMENTING WITH POSSIBILITIES

Imagine a society where language doesn't allow people to speak about the future. There are no words for "tomorrow," "later," or "next year." Future-tense verbs don't exist. Instead, all events must be described in the present or the past. Climate change, in this world, can only be discussed in terms of what has already happened, such as droughts, wildfires, typhoons, or floods, but never in terms of what *might* come. Without a linguistic structure to frame future consequences, how would people plan to take action? Would urgency fade without the ability to articulate the long term? Or would it be part of everyday responsibility? If language shapes perception through psychological time, could changing the way we speak about time change how we act in the present?

End of the world (as we know it)

Here's a bold statement that we're just going to put out there as both a coda and extension of this book: it's the end of the world (as we know it).

The band REM's song of the 1980s already popularized this notion, with the ironically apathetic "and I feel fine" to conclude the phrase. Although not news to some people, it's relevant now more than perhaps it was even during the Cold War 1980s period. This might literally mean the extinction of many living species on the planet, including humans.

Another way to consider this more broadly is to reflect: is the *end* such a bad thing? Such a question may sound somewhat cheeky, but bear with us a moment. *Where do we end?* begs the question of *how do we end?* – a transitional invitation that has cyclical possibilities related to *where do we begin … again?* Is the nature of time linear, circular, or perhaps spiral, like Aristotle's *golden mean*, an ethics position that virtues are to be found by balancing extreme points of view?

Much of the transition we're experiencing is the final chapter of an older, materialistic world built on economies of consumption and waste that contained a different set of values and priorities. This older, dying world has often been referred to as *modernity*, which is characterized by a shift from traditional, agrarian societies to urban "enlightened" and "industrialized" societies dating back as early as the late fifteenth century and up to the present, a period that has given rise to the Anthropocene.

The success of modernity is deeply intertwined with the languages and narratives that shaped it. Phrases like Descartes's "I think, therefore I am," Galileo's "nature is a book written in mathematical language," Benjamin Franklin's "time is money," and Ayn Rand's "civilization is the progress toward a society of privacy" reflect the evolving worldview of modernity. This worldview emphasizes an overreliance on cognitive reasoning, scientific understanding, efficiency, and hyper-individualism.

————ooOoo————

In her powerful book *Hospicing Modernity*, educational leader Vanessa Machado de Oliveira Andreotti uses the metaphor of "the house that modernity built" to show how our ways of thinking, speaking, and acting, as elements of discourse, are shaped by dominant systems of power.[2] Language continues to play a huge role in keeping this house of modernity standing. Words like "progress" and "development" make it seem like there's only one right way to live, usually based on Western ideas of endless growth. When people call a country "developing," for instance, it suggests it's lacking something rather than simply being different. If we change the way we speak – using terms like "regeneration" instead of "development" or focusing on "coexistence" rather than "endless growth" – we can start imagining new ways of living that don't just repeat the same old patterns. Rather than living in the same old crumbling house, expecting it will continue to support us, the aim is to co-create a new community as the *living spaces that ecology built.*

While the historical period of modernity has varying degrees of benefits and drawbacks, it nevertheless has dominated global worldviews and practices for well over 300 years. A different world, one that's still being shaped, is the future, and it's happening right now.

The transitional period of one or multiple worlds to others, as sociologist Hartmut Rosa considers, constitutes the always present anxiety of modernity. This is a crucial point. The anxieties we face – whether fear of ecological collapse on one end of the spectrum or longing for an industrial past on the other end – arise from the loss of a familiar world of modern ideals and the uneasy transition into another world still unfolding.

As Rosa reminds us, one of the hallmarks of modernity is *certainty*, or the illusion that we can control people, institutions, or even the outcome of events through things like predictability and efficiency when, in fact, the world has always been uncontrollable and inherently uncertain, shifting across various historical paradigms.[3]

The end of one dominant worldview and power structure could be another way of saying *transition*, which, like futuring, is how we can make sense of both biological and psychological time through the varying degrees of uncertainty. While unpredictability of our planet's future looms, major shifts are happening in energy and the economy as society confronts ecological and climate disasters.

The flaws of an economic system built on endless unsustainable growth, especially when that growth is hitting its planetary limits, is becoming too obvious to ignore. There needs to be a time of transformation and regeneration for anything to be sustainable, like a seasonal garden moving through natural cycles of growing and withering away.

So, by way of conclusion, that brings us to two uncertain realities about two massively important unknowns: our economic and energy futures. Often called the *energy transition*, this term refers to the global shift from fossil fuels like coal, oil, and natural gas to cleaner, more sustainable energy sources like solar, wind, hydro, biomass, and other renewable energy sources. Such a transition involves much more than simply swapping out old technologies for new ones. It also requires finding new language for how we live, so we can move beyond systems built on perpetual growth and destructive waste like emissions.[4]

Let's be clear, though, this shift has already begun, arguably in the latter half of the twentieth century, when renewable forms of energy were widely recognized as essential for future survival. Still, there's a risk in treating transition as a completed task, when in fact we're continuously *energy transitioning*. If we hold onto inherited ways of thinking and speaking and acting, we put the whole planet's ecological systems at risk.

Because the energy transition isn't just about swapping out fossil fuels for renewables, it must also account for political and social shifts that touch on people's lives, affecting their values and beliefs. What fuels sociopolitical change? How we talk about energy and sustainable futures shapes public perception, policy decisions, economic interventions, and collective action. Engagements with this transition often overlook a crucial element: the role of language as an influence in how we live and make decisions.

Consider the difference between saying "phasing out fossil fuels" versus "investing in clean energy." The former emphasizes loss and resistance that may come from it, while the latter frames the transition as an opportunity. Countries like Germany use terms like *Energiewende* (literally "energy turn") to signal a national commitment to a renewable future, making the transition feel intentional and structured. Whereas in North America, the language about energy often focuses on "energy security" or "energy independence," often tied to freedom and patriotism, which can reinforce fossil fuel reliance rather than emphasize transformation.

Beyond policy, everyday language also matters. When people hear "net zero" or "carbon-neutral," the terms may feel abstract and divorced from people's daily needs. Or, even worse, the term "decarbonized oil," which ignores the fact that oil is about 85 percent carbon. Even the term *clean coal* is a misnomer for obvious reasons. These examples, among many others, represent a classic discursive "reframing" of language to alter people's perception. But framing energy shifts around concepts like "clean air for our kids" or "affordable energy for all" can make the transition more relatable without relying on terms that mislead meaning.

> The energy transition is as much about language
> and meaning as it is about policy or technology.

Even the word *transition* suggests a major shift, but it specifically highlights a gradual, large-scale change unfolding over generations. It gives the initial impression that the process is linear, like a clear path from one point to another, neatly divided into stages. In truth, the reality is far messier. Transitioning is less like crossing a bridge and more like navigating a winding river, encountering unforeseen obstacles and making course corrections along the way.

As Canadian energy futurist Imre Szeman asks in his book *Futures of the Sun: Struggle Over Renewable Life*, "who will lead the transition from fossil fuel-dependent societies into renewable energy futures?"[5] It's a question that highlights the need for different leadership and collective management, much like how movements of Afrofuturism or Indigenous futurism consider other possibilities for how futures can be reimagined through alternative discourses that affect multiple worldviews. As we can now appreciate, the language we use influences how people understand and engage with change. In that sense, we're all part of the transition, influencing its direction through the dialogue we have every day.

The finance sector is also undergoing a transition toward sustainability, with rising awareness in "green finance" and "socially responsible investing." Terms like "ESG (environmental, social, and governance) criteria," "impact investing," and "green bonds" are becoming standard as investors seek to align their financial interests with environmental and social goals. While we may not be advocating for green finance based upon the current capitalist systems per se, these examples still show how terminologies have led to different conceptions of "sustainable" approaches to finance and economics.

Moving even further toward transformation, the concept of a *circular economy*, where resources are used and reused in a closed loop rather than discarded after a single use, is gaining traction in various cities and companies around the world. Patagonia's Worn Wear program provides a successful example of circular economy. By repairing, refurbishing, and reselling used clothing, this program extends a ski coat's life cycle while also promoting sustainable consumption for consumers.

Because of business models like the outdoor gear company Patagonia, terms like "circularity," "resource efficiency," and "waste reduction" are becoming more common as businesses and policymakers explore ways to transition toward more sustainable production and consumption models. Circular economies, while appearing unlikely for entire societies in the near future, present functional models for living more ecologically.

Circular or regenerative approaches to the economy have significantly altered language and meaning surrounding sustainability. It has shifted the focus from a linear *take-make-dispose* model to one that emphasizes the regenerative use of resources, including waste reduction and the promotion of sustainable production and consumption practices. This change in discourse has prompted conversations on reimagining business models, including entire economic systems, designing products for longevity and reuse, encouraging innovation in recycling and resource recovery such as upcycling, and promoting collaboration across various industries. With these shifts in both perception and practice come further changes in language and communication.

Circularity implies that we're constantly engaging with the world through dialogue, and this is what becoming ecological is all about. Although this all may seem like day-to-day stuff, these examples show how futuring different worlds in the present works, largely through shifting language and communication that eventually shapes meaning through action. If futuring is a state of transition, combining past-present-future all at once, then these commonplace changes can have large impact.

While examples of shifting from single-use economies, where waste is merely a by-product of economic progress, or even more progressive circular economic models, the *end of the world* also involves shifting worldviews and entire social and political systems. While this may sound revolutionary, it's actually pretty straightforward, and it's already happening. By gradually shifting culture and society in ways that create a fairer, more sustainable world, the end of the old world becomes a transition into a different one, not an end to all.

PRACTICING

In a group or on your own, create a Worlds in Transition map. Draw a large circle on the map and divide it into four quadrants. Label each section: 1) "Worlds Ending" – things that are disappearing or need to; 2) "Worlds Emerging" – new systems, technologies, or ideas shaping the future; 3) "Narratives of Loss" – stories we tell about decline, collapse, or fear; and 4) "Narratives of Possibility" – stories of hope, transformation, or new ways of being. Use images or short descriptions to fill in each section. Afterward, dialogue or reflect on these questions: how do the narratives we use shape our language and actions? How might shifting our language help us move from fear to possibility? How does our participation in ongoing transitions help facilitate change?

Possible futures

Because of the tendency for humans to create fear out of the unknown, the future is often the most focused-on aspect of time. The future might often be associated with fear, but it also holds immense possibility and even opportunity through imaginative capacities for change. Put out as an invitation, let's see what the future our tea leaves reveal before jumping to worst-case scenarios.

When cultivating these spaces of curiosity and openness, we encourage ways to build responsible futures that recognize the value of all life, embedding these principles in our discourses through practices such as deep listening, witnessing various stories and experiences, and unsettling the cultures around us through the dynamic process of living in relation with each other. Much of the fear we might be feeling is one of transition. In this transition, we're experiencing a death of an outdated, materialistic, and modern world that revolved around a grand narrative of economic progress and growth with its attendant set of unsustainable values and priorities.

When we look closely, transition is all about the *in-betweenness* of possible worlds. The word itself comes from the Latin *transitio*, meaning "a going across or over," from *transire*: *trans* ("across") + *ire* ("to go"). It's a word rooted in motion and doing, also in the verb-like quality of becoming. Transition is not a fixed point but a crossing as an active process of moving through uncertainty, like a river winding between banks of meaning.

But that "in-between" is not just a pause or a bridge between fixed points. It's an active, unfolding space where previously rooted realities are uprooted and re-formed momentarily and then subject to continued change. Points aren't fixed but dynamically fluid. Transition, in this sense, isn't so much arriving somewhere else as it is about what gets shaped and reshaped along the way as an ongoing process of becoming ecological.

What's more, the in-betweenness embedded in the term *transition* holds both tension and potential. It captures the threshold space where things are often unsettled and uncertain, but also full of futuring possibility. Such an experience of living invites experimentation and with it often discomfort because it requires letting go of the known before the new is fully formed.

In times of crisis and change, paying attention to this transitional space can help us notice what's shifting, what's emerging, and possibly what kinds of futures we might cocreate in the process. Futuring embraces a different world composed of many small stories and collaborative actions taken to preserve what's left of the ecological present and imagine a different world, one that is created through acknowledging our interdependence on the Earth, always emerging, often mutable and diffuse, and always interconnected. In these futures that can be scaled beyond humanity alone, a dialogue must emerge that includes all living beings.

As Belgian physical chemist Ilya Prigogine wrote in his aptly titled book *The End of Certainty*, "The future is under perpetual construction,"[6] by which we could say it's *always in the making* similar to what we mean by *becoming ecological*. What the future will be for us personally and socially remains a complex web of possibilities that although exist in the present have consequences that amass over time. We emphasize possibilities to point out that negative future visions aren't inevitable if we become conscious, weighing personal desire and convenience against the entire network of relations, both human and planetary.

The ongoing process of becoming ecological presents myriad opportunities for "constructing" many possible futures through the ways we all talk, think, and act through our ever-present discourse ecologies. What you choose to do in the next five minutes, hour, or day could ripple out and influence people and outcomes in ways you can't fully predict.

As a global society, we're still in a position to become conscious, attending to various possible choices, which is where our optimism grows. After all, there's no end (of the world), only transition, crossing the unknown. This is perhaps one of the most profound connections among ecology, language, and meaning.

We also make choices every time we communicate. How we express our needs and desires, and how we engage with and learn from one another, will be the result of the language that influences society at large. Because the future is ecological, much like the present, it depends upon the relational elements of co-creation to make any meaning out of the experience. The outcomes of whether an old-growth forest will be logged and harvested depends upon other people, and the future of this forest is part of a collective responsibility. In this sense, futuring isn't individual, or about humans, but about being in relational dialogue with all living beings on the planet, always in transition.

As a way to conclude by circling back to the beginning, we are reminded that *we compose our worlds through words*. Quite literally. Language is a social process of composition, not just using words, but how meaning is made. To compose is to form futures, inviting people into imaginative possibilities. We continue to create the worlds that we live in. Just as every day is a new day, we can always create different futures in each moment. Meaning exists through the combined actions of groups of people, and we are meaning-making animals. That's what we humans do.

> Language, and communication in its entirety, is one of the richest resources we still have in the effort to stabilize the future.

Embracing and supporting diverse dialogue, holding multiple viewpoints, and creating possible futures assists the effort to adapt and (hopefully) evolve human behavior. This is true whether we aim to make a more socially just world or to relationally build a more ecologically livable world, to hit the brakes on the runaway catastrophe train and give us humans a reasonable chance to turn down the temperature of the overheating world.

Futuring, then, is the ultimate discursive ecology because it simultaneously deconstructs pasts, composes possible futures, and enacts them in the present moment. The magical experience of living and being alive embraces futuring. Actively becoming conscious in this process changes its impact, reverberating across society.

Starting now, let's keep futuring by tending to the language, growing the dialogue and continuing our process of becoming ecological.

REFLECTING

- How does the tension between fear and possibility influence the language we use to envision and shape sustainable futures?
- What are some additional terms, beyond "net-zero," "green finance," and "circular economy," that could help shape public perception and drive collective action toward a more sustainable and equitable future?
- How does the language used to describe energy transitions influence public perceptions, policy decisions, and collective action in shaping ecologically sustainable futures? What about transition more generally?

Group Dialogue Questions

1. Based on the definition of discursive ecologies in **Chapter 1, Discoursing**, what are some examples in your daily lives where language has created meaning and influenced your worldviews around ecological matters? How often do you consider how the language you use creates meaning and supports your worldviews or belief systems. How do you see these as interconnected with the ecological world, such as how things interconnect with people or even the more-than-humans around you like trees, soil, birds, bees, and others?

2. Looking at examples of being and becoming in **Chapter 2, Becoming**, how do you see yourself becoming ecological over time (consider pasts as well as futures)? When reflecting on your language practices, how might you engage in dialogue with those around you? If becoming looks at people as in a

state of always arriving, adapting, and changing, then how does the process and practice of becoming look like even after reading this book?

3. **Chapter 3, Listening**, introduces the idea that deep and collective listening is a core practice of becoming ecological. In what ways do you use selective listening to avoid taking action and changing beliefs? What are the sounds that you connect with most deeply when reflecting on your relationship to nature? Listening requires us to slow down time from the rapid speed of visual information. Deep listening engages with timeframes that are slower and lower than audible sound. How might we slow down our sense of time to be able to hear the planet speaking with us?

4. In **Chapter 4, Relating**, concepts of individual and relational perspectives present two quite different paradigms. How might our sense of the "individual" or "self" inhibit environmental awareness and action? What are some examples from your own culture that demonstrate relational approaches to living ecologically? How might other cultures also portray greater value on socioecological relationships? What parts of your life could feel less about "I" or "self" and more about co-creating meaning with the people around you? Ecological relating also moves beyond just human relationships. In what ways do you relate to the more-than-human world?

5. **Chapter 5, Ingesting**, places a special focus on benefits and challenges related to convenience and fast food versus and slow food. Ingesting also concerns what

and how we ingest and digest language and meaning. Consider the old saying "you are what you eat." Do you know very much about the food you eat? Where it comes from, how it was prepared, and what was put into it? Now consider the information diet you consume. Do you know where it comes from, who made it, or what agendas have been injected into it? How have these experiences changed the way you view the world? What about the ecological world?

6. In **Chapter 6, Composting**, we bring to light the importance of regenerative cycles in sustaining all life, and the problem of offcycling, which is the origin of waste and pollution. In what way does excess and waste support your lifestyle and identity? How does waste make life more challenging, and how do you address these contradictions? Can you think of any ways to reuse old memories and upcycle old things, so that they become more useful or valuable than they were before? How are regenerative cycles present in discourse? How do we take up ecological imperatives like recycle, reduce, and reuse in language?

7. In Chapter 7 we explore the notion of **Unsettling**, in both a positive and negative light. We propose that as a part of unsettling some habitats to prevent ecological collapse and mass extinction, we need to unsettle our habits of thinking arising from habitual and predictable language use. Anomalies, or irregular patterns, are key to identifying when new models are needed. Have you had an experience where something completely unpredicted occurred and

caused you to change your mind or your habits about the environment? Has your knowledge caused you to settle intellectual territories of thought that resists change? What would it take to make you unsettle those territories, to migrate your thinking?

8. **Chapter 8, Storying,** considers how narratives and stories present pathways for social and ecological change. How were stories some of the most influential teachings that ultimately changed your mind about something? What are some examples of ecological or climate stories that you have been exposed to? How did they affect your views? Storying engages us as listeners, but it can also compel us to recount our own stories. What stories would you tell about the environment to invite others to embrace new ways of embracing becoming ecological?

9. In **Chapter 9, Witnessing,** explores the relationship between witnessing and giving testimony as a function of ecological discourse that has the power to motivate profound change. We put attention on witnessing two kinds of catastrophic events: commonplace catastrophe, which many older people bear witness to everyday, and celebrity catastrophe, which is primarily a mediated form of witnessing cataclysmic events. Can you think of any commonplace or celebrity catastrophes that greatly influence your behavior and outlook on the world. How has this made you more ecologically aware? Have you told this story to other family or friends or perhaps even on social media?

10. The book presents an invitation to actively collaborate on building different futures, as outlined in **Chapter 10, Futuring**. How often do you think about the future? How much of this do you actively envision different futures? If futuring is a language of practice, then how might language contribute to a creation of different possible worlds or worldviews, ones that align with ecological sustainability practices? In what ways, and through which stories, do we explore generative, sustainable futures? How might your language around concepts like "time" or "transition" shift, knowing that it directly shapes how we make meaning and imagine other possible futures?

Acknowledgments

We acknowledge that we are part of the larger planetary system and appreciate this moment to engage in the ongoing dialogue with all of the other human and more-than-human voices that make sharing life possible. We also recognize that language and ecology books emerge out of systems of culture, ideas, experiences, histories, cosmologies, and so on. Many people have been part of this journey – directly and indirectly, future and past – and we are deeply thankful for the opportunity to engage with these ideas and teachings.

Books also require publishing support. For that, we are grateful to Aevo and the team at the University of Toronto Press. A special thanks goes out to Jodi Lewchuk, our acquisitions editor, for believing in this project from the very beginning. Thanks for visioning with us and making this book a reality.

Notes

Prefacing

1 Jiddu Krishnamurti, *Facing A World in Crisis: What Life Teaches us in Challenging Times*, ed. David Skitt (Boston: Shambala, 2005).
2 For more on depth education, see Vanessa Machado de Oliveira, *Hospicing Modernity: Facing Humanity's Wrong and the Implications for Social Activism* (Berkeley: North Atlantic Books, 2021).
3 Robin Wall Kimmerer, "Speaking of Nature: Finding Language That Affirms Our Kinship with the Natural World," *Orion Magazine*, 12 June, 2017, https://orionmagazine.org/article/speaking-of-nature/.
4 See Bruno Latour, *How to Inhabit the Earth: Interviews with Nicolas Truong*, trans. Julie Rose (Cambridge, UK: Polity, 2024).
5 Instead of using the term "non-human," which centers humans as the default, we draw on "more-than-human," a phrase introduced by David Abram in *The Spell of the Sensuous: Perception and Language in a More Than Human World* (New York: Pantheon, 1996), to better reflect the dynamic agency of all living beings.

Chapter 1 – Discoursing

1 See Elizabeth Kolbert, *The Sixth Extinction: An Unnatural History* (New York: Picador, 2015), and Timothy Morton, *Being Ecological* (New York: Penguin, 2018).
2 Others have also explored helpful ways of approaching living and being ecological. For starters, see Robin Wall Kimmerer, *Braiding

Sweetgrass: Indigenous Wisdom, Scientific Knowledge, and the Teachings of Plants (Minneapolis, MN: Milkweed, 2013), Timothy Morton, *Being Ecological* (New York: Penguin, 2018), and Andreas Weber, *Enlivenment: Toward a Poetics for the Anthropocene* (Boston: MIT Press, 2019). Morton's book wittingly challenges conventional ideas of what it means to "be" ecological and, while different in many respects, shares a similar spirit with our approach, especially through the titles of both books.

3 Wallace Stevens, *The Snow Man*, in *Harmonium* (New York: Alfred A. Knopf, 1923). See also James W. Heisig, *Much Ado About Nothingness: Essay on Nishida and Tanabe* (Nagoya, Japan: Chisokudō Publications, 2016), 13.

4 Bruno Latour, "Why Has Social Critique Run Out of Steam? From Matters of Fact to Matters of Concern," *Critical Inquiry* 30 (2004): 225–48, www .bruno-latour.fr/sites/default/files/89-CRITICAL-INQUIRY-GB.pdf.

5 While linguists have made important contributions to studying language, linking it to ecology remains a niche area of study. Fortunately, the interdisciplinary academic field of "ecolinguistics" bridges this gap, exploring the deep connections between language and the environment. See, for instance, Arran Stibbe, *Ecolinguistics: Language, Ecology and the Stories We Live By*, 2nd ed. (New York: Routledge, 2021).

6 Charles Taylor, *The Language Animal: The Full Shape of the Human Linguistic Capacity* (Cambridge, MA: Belknap Press of Harvard University Press, 2016).

7 Chris Hatch, "Should we even bother talking about climate change?" *National Observer*, 18 February 2025, www.nationalobserver.com/2025/02 /14/opinion/should-we-even-bother-talking-about-climate-change

8 Ursula K. Le Guin, "Acceptance Speech for the Medal for Distinguished Contribution to American Letters," *National Book Foundation*, 19 November 2014, www.nationalbook.org.

9 Mark Twain, Letter to George Bainton (1888), www.twainquotes.com /Lightning.html. Italics our emphasis

10 Kinji Imanishi, *The World of Living Things*, trans. P.J. Asqith, H. Kawakatsu, S. Yagi, & H. Takasaki (London: Routledge, 2002).

11 Kamyar Razavi, "Animals and Purity: Ways to Influence Trump 2.0 on the Environment," *Medium, The New Climate*, 29 January 2025, https:// medium.com/the-new-climate/animals-and-purity-ways-to-influence -trump-2-0-on-the-environment-7c078175faad.

12 Vyvyan Evans, *The Crucible of Language: How Language and Mind Create Meaning* (Cambridge: Cambridge University Press, 2015), 54.

13 Dougald Hine, *At Work in the Ruins: Finding Our Place in the Time of Climate Crises and other Emergencies* (London: Chelsea Green, 2023).

14 In simplifying discourse for everyday understanding and its connection to ecology, we couldn't provide an in-depth analysis in this book. For readers seeking a deeper exploration of discourse, we recommend the following resources: Judith Butler, *Excitable Speech: A Politics of the Performative* (New York: Routledge, 1997); Jacques Derrida, *Writing and Difference*, trans. Alan Bass (Chicago: University of Chicago Press, 1978; original work published

1967); Michel Foucault, *The Archaeology of Knowledge*, trans. A. M. Sheridan Smith (New York: Pantheon Books, 1972); Deborah Tannen, *Talking Voices: Repetition, Dialogue, and Imagery in Conversational Discourse*, 2nd ed. (New York: Cambridge University Press, 2007).

15 As quoted in Byung-Chul Han, *The Philosophy of Zen Buddhism*, trans. Daniel Steuer (London: Polity, 2022), 59.

16 Bill McKibben, "A Smoking Gun for Biden's Big Climate Decision," *The New Yorker*, 31 October 2023, www.newyorker.com/news/daily-comment /a-smoking-gun-for-bidens-big-climate-decision.

17 Alan Watts, *The Way of Zen* (New York: Vintage, 2019).

18 Aldo Leopold, *A Sand County Almanac* (New York: Oxford University Press, 1949).

19 Yoshifumi Miyazaki, *Shinrin Yoku: The Japanese Art of Forest Bathing* (Portland, OR: Timber Press, 2018).

20 Marcia Bjornerud, *Timefulness: How Thinking Like a Geologist Can Save the World* (Princeton: Princeton University Press, 2018).

21 Vyvyan Evans, *The Crucible of Language: How Language and Mind Create Meaning* (Cambridge: Cambridge University Press, 2015), 7.

22 Buckminster Fuller and Quentin Fiore, *I Seem to Be a Verb* (New York: Bantam, 1970).

23 WWF, *Living Planet Report 2024 – A System in Peril* (Gland, Switzerland: WWF, 2024), https://livingplanet.panda.org/en-US/.

24 Neil Theise, *Notes on Complexity: A Scientific Theory of Connection, Consciousness, and Being* (New York: Spiegel and Gran, 2023), 38. See also Annalee Newitz, *Scatter, Adapt, and Remember: How Humans Will Survive a Mass Extinction* (New York: Viking, 2013) for ways of considering mass extinction as a transition, one of many over the years on the planet, not only as an end.

25 Humberto R. Maturana and Francisco J. Varela, *Autopoiesis and Cognition: The Realization of the Living* (Dordrecht: D. Reidel Publishing Company, 1980).

26 Suzanne Simard, *Finding the Mother Tree: Discovering the Wisdom of the Forest* (London: Allen Lane, 2021).

27 Thomas F. Homer-Dixon, *The Upside of Down: Catastrophe, Creativity, and the Renewal of Civilization* (Washington, DC: Island Press, 2006); see the term *catagenesis* on p. 289.

28 Edgar Morin, *Homeland Earth: A Manifesto for the New Millenium*, trans. Anne Brigitte Kern (Cresskill, NJ: Hampton Press, 1999).

Chapter 2 – Becoming

1 See Timothy Morton, *Being Ecological* (New York: Penguin, 2018).

2 Kitarō Nishida, *Intelligibility and the Philosophy of Nothingness: Three Philosophical Essays*, trans. Robert Schinzinger (Pantianos Classics, 1958).

3 Roger T. Ames, "What Ever Happened to 'Wisdom'? Confucian Philosophy of Process and 'Human Becoming'," *Asia Major* 21, no. 1 (2008): 45.

4 Kenneth Gergen, *Relational Being: Beyond Self and Community* (Oxford: Oxford University Press, 2009).

5 Thich Nhat Hanh, *Interbeing: Fourteen Guidelines for Engaged Buddhism* (Berkeley, CA: Parallax Press, 1987)

6 Paulo Freire, *Pedagogy of the Oppressed*, trans. Bergman Ramos (New York: Continuum, 2005).

7 Donna Haraway, *When Species Meet* (Minneapolis: University of Minnesota Press, 2008).

8 David Bohm, *On Dialogue* (New York: Routledge, 1996). See also Kenneth Gergen, *An Invitation to Social Construction: Co-Creating the Future*, 4th ed. (London: Sage, 2023).

9 Gergen, *An Invitation to Social Construction*.

10 Ludwig Wittgenstein, *Philosophical Investigations* (Oxford: Blackwell, 1953). See also Gergen, *An Invitation to Social Construction*, for a contextual explanation of Wittgenstein's work.

Chapter 3 – Listening

1 *First Peoples Principles of Learning* (First Nations Education Steering Committee, 2007), www.fnesc.ca/first-peoples-principles-of-learning/.

2 Don Hill, "Listening to Stones: Learning in Leroy Little Bear's Laboratory – Dialogue in the World Outside," *Albertaviews: The Magazine for Engaged Citizens*, 1 September 2008, https://albertaviews.ca/listening-to-stones/.

3 Graham Parkes, "Kūkai and Dōgen as Exemplars of Ecological Engagement," in *Japanese Environmental Philosophy*, ed. J. Baird Callicott and James McRae (New York: Oxford University Press, 2017), 65–86.

4 C.N. Gamble, J.S. Hanan, and T. Nail, "What Is New Materialism?" *Angelaki* 24, no. 6 (2019): 111–34, https://doi.org/10.1080/0969725X.2019.1684704. See also additional resources at https://philosophy-of-movement.com/tag/material-ecology/.

5 Corryn Wetzel, "The Pandemic Shutdown in San Francisco Had Sparrows Singing Sexier Tunes," *Smithsonian Magazine*, 25 September 2020, www.smithsonianmag.com/science-nature/pandemic-shutdown-san-francisco-had-sparrows-singing-sexier-tunes-180975913/. See also the *World Soundscape Project* at Simon Fraser University and its findings over the past 50+ years, www.sfu.ca/~truax/wsp.html.

6 In addition to matters of fact or concern, as seen in Bruno Latour, "Why Has Social Critique Run Out of Steam?: From Matters of Fact to Matters of Concern," *Critical Inquiry* 30 (2004): 225–48, www.bruno-latour.fr/sites/default/files/89-CRITICAL-INQUIRY-GB.pdf, also look at Isabelle Stengers, *Another Science is Possible: A Manifesto for Slow Science*, trans. Stephen Muecke (Cambridge, UK: Polity, 2018).

7 Andreas Weber, *Enlivenment: Toward a Poetics for the Anthropocene* (Boston: MIT Press, 2019).

8 Walter J. Ong, *Orality and Literacy: The Technologizing of the World* (1982; reprint: Routledge, 2012).

9 See, for instance, Matthew Tomlinson, "'Clean Communication': Felt-Sense Methodologies and the Reflexive Researcher in Equine-Assisted Personal Development," *The Sociological Review* 72, no. 1 (2024): 155–74, https://doi.org/10.1177/00380261231186752.

10 "Compassion Fatigue: Signs, Symptoms, and How to Cope," *Physician Wellness Hub, Canadian Medical Association,* 8 December 2020, www.cma.ca/physician-wellness-hub/content/compassion-fatigue-signs-symptoms-and-how-cope.

Chapter 4 – Relating

1 Lev Vygotsky, "The Genesis of Higher Mental Functions," in *The Concept of Activity in Soviet Psychology*, ed. J.V. Wertsch (Armonk, NY: M.E. Sharpe, 1981).

2 Melanie Goodchild, "Relational Systems Thinking: The Dibaajimowin (Story) of Re-theorizing 'Systems Thinking' and 'Complexity Science'," *Journal of Awareness-Based Systems Change* 2, no. 1 (2022): 53–76. https://doi.org/10.47061/jabsc.v2i1.2027.

3 For another critical viewpoint of the Enlightenment, see David Graeber and David Wengrow, *The Dawn of Everything: A New History of Humanity* (Toronto: Signal McClelland & Stewart, 2021). In this book, they reveal how some Enlightenment thinkers were influence by Indigenous thinkers in the "new world," even though these relationships were not openly acknowledged.

4 René Descartes, *Discourse on Method* (New York: Macmillan, 1986).

5 Philippe Descola, "Ta pensée audacieuse est devenue la pensée du temps present," *Le Monde*, 3 November 2022.

6 Marshall McLuhan and Quentin Fiore, *The Medium is the Massage* (New York: Penguin, 1967).

7 Questions of the "self" run deep and might need further exploration. We recommend some of the following books. Jay L. Garfield, *Losing Ourselves: Learning to Live Without a Self* (Princeton: Princeton University Press, 2022) and Mark Siderits, Evan Thompson, and Dan Zahavi, eds., *Self, No Self? Perspectives from Analytical, Phenomenological, & Indian Traditions* (New York: Oxford University Press, 2013).

8 For more on this, see Kitarō Nishida, *An Inquiry into the Good*, trans. By M Abe and C. Ives (New Haven, CT: Yale University Press, 1990).

9 These words are spoken in the opening of the song "Your Ex Lover Is Dead" by the Canadian band Stars. See Stars, *Set Yourself on Fire* (Arts & Crafts, 2004).

10 Sheila McNamee and Kenneth J. Gergen, *Relational Responsibility: Resources for Sustainable Dialogue* (New York: Sage, 1999); Kenneth Gergen, *Relational Being: Beyond Self and Community* (New York: Oxford University Press, 2009).

11 Robin Wall Kimmerer, *Braiding Sweetgrass: Indigenous Wisdom, Scientific Knowledge, and the Teachings of Plants* (Minneapolis: Milkweed Editions, 2013), 48–9.

12 John Urry, *Societies Beyond Oil: Oil Dregs and Social Futures* (London: Zed Books, 2013).

13 Patricia M. Greenfield, "The Changing Psychology of Culture From 1800 Through 2000," *Psychological Science* 24, no. 9 (2013): 1722–31, https://doi .org/10.1177/0956797613479387.

14 English is the most spoken language as of 2025, which includes both native and nonnative speakers. Mandarin Chinese has the most native speakers out of any language at this point.

15 Robin Wall Kimmerer, *Braiding Sweetgrass: Indigenous Wisdom, Scientific Knowledge, and the Teachings of Plants* (Minneapolis, MN: Milkweed, 2013), 53.

16 Kimmerer, *Braiding Sweetgrass*, 57.

17 Tom Oliver, *The Self Delusion: The Surprising Science of Our Connection to Each Other and the Natural World* (London: Weidenfeld & Nicolson, 2020), 153.

18 Robin Wall Kimmerer, "Speaking of Nature: Finding Language That Affirms Our Kinship with the Natural World," *Orion Magazine*, 12 June 2017, https://orionmagazine.org/article/speaking-of-nature/.

19 Gavin Van Horn, "Kinning: Introducing the Kinship Series," in *Kinship: Belonging in a World of Relations, Vol. 1 Practice*, ed. Gavin Van Horn, Robin Wall Kimmerer, and John Hausdoerffer (Libertyville, IL: Center for Humans and Nature, 2021), 3. This six-volume series explores the dynamic process of kinning in multiple ways, highlighting its active, verb-like nature.

20 Kyle P. Whyte, "An Ethic of Kinship," in *Kinship: Belonging in a World of Relations, Vol. 5 Practice*, ed. Gavin Van Horn, Robin Wall Kimmerer, and John Hausdoerffer (Libertyville, IL: Center for Humans and Nature, 2021), 35.

21 Kimmerer, *Braiding Sweetgrass*, 56.

22 Richard Wagamese, "Living with Bears," *Yukon News*, 21 August 2008, www.yukon-news.com/news/living-with-bears/.

23 Enrique Salmón, "Kincentric Ecology: Indigenous Perceptions of the Human-Nature Relationship," *Ecological Applications* 10, no. 5 (2000): 1327–32.

24 It's worth noting here, for those wishing to dive deeper into this topic, how the philosopher duo of Gilles Deleuze and Felix Guattari both emphasized the ontological reality of "becomings," as we coexist not just as separate beings but also integral becomings. See Gilles Deleuze and Félix Guattari, *A Thousand Plateaus: Capitalism and Schizophrenia*, trans. Brian Massumi (Minneapolis: University of Minnesota Press, 1987).

25 Jacob Richey and Paul Wapner, "The Inner and Outer Ecologies: Contemplative Practice in an Environmental Age," *The Arrow: A Journal of Wakeful Society, Culture & Politics* 4, no. 1, https://arrow-journal.org/inner -and-outer-ecologies-contemplative-practice-in-an-environmental-age/.

Chapter 5 – Ingesting

1 Paul Virilio, *The Information Bomb*, trans. Chris Turner (London: Verso, 2005).
2 William Blake, *The Marriage of Heaven and Hell* (Boston: John W. Luce and Company, 1906), www.gutenberg.org/files/45315/45315-h/45315-h.htm.
3 "The Swanson 'TV Dinner,' Which Hit Grocery Store Cases on September 10, 1953, Was an Immediate Success. In 1954, Swanson Sold More Than 10 Million Units, and the Next Year, 25 Million," *History*, www.history.com/news /tv-dinner-history-inventor.
4 David F. Noble, *Forces of Production: A Social History of Automation* (New Brunswick: Transaction Publishers, 2011), https://deterritorialinvestigations .files.wordpress.com/2015/03/david_f-_noble_david_f-_noble_forces_of _productbookza-org.pdf.
5 Ted T. Aoki, "Legitimating Lived Curriculum: Toward a Curricular Landscape of Multiplicity," *Journal of Curriculum and Supervision* 8, no. 3 (1993): 255–68, https://acurriculumjourney.wordpress.com/wp-content /uploads/2014/04/aoki-1993-legitimating-lived-curriculum-towards-a -curricular-landscape-of-multiplicity.pdf.

Chapter 6 – Composting

1 Susan Strasser, *The Other Side of Consumption: Waste and Want* (Oxford: Berg, 1992).
2 Julian Jaynes, *The Breakdown of the Bicameral Mind* (New York: Mariner Books, 2000).
3 Thorstein Veblen, *The Theory of the Leisure Class: An Economic Study of the Establishment of Institutions* (New York: Macmillan, 1899).
4 Annie Leonard, *The Story of Stuff: How Our Obsession with Stuff Is Trashing the Planet, Our Communities, and Our Health – and a Vision for Change* (New York: Free Press, 2010).
5 Joseph Roach, *Cities of the Dead: Circum-Atlantic Performance* (New York: Columbia University Press, 2021), 41.
6 R. Marfella et al., "Microplastics and Nanoplastics in Atheromas and Cardiovascular Events," *The New England Journal of Medicine* 390, no. 10 (2024): 900–10, www.nejm.org/doi/full/10.1056/NEJMoa2309822.
7 Mikhail M. Bakhtin, *Art and Answerability: Early Philosophical Essays*, trans. Kenneth Brostrom, ed. Michael Holquist (Austin: University of Texas Press, 1990).
8 See, for instance, Jonathan Haidt, *The Anxious Generation: How the Great Rewiring of Childhood Is Causing an Epidemic of Mental Illness* (New York: Penguin, 2024).
9 E.F. Schumacher, *Small Is Beautiful: A Study of Economics as if People Mattered* (London: Vintage, 1973).

Chapter 7 – Unsettling

1 From Narrow to General AI, "Experiences Can't Contradict Each Other, Only Motives Can," *Medium*, 12 April 2024, https://ykulbashian.medium.com/experiences-cant-contradict-each-other-only-motives-can-0bf3ea13cf88.
2 Viktor Schauberger, *The Water Wizard: The Extraordinary Properties of Natural Water*, trans. Callum Coats (London: Gill Books, 1999).
3 Ivan Illich, *H2O and the Waters of Forgetfulness: Reflections on the Historicity of "Stuff"* (London: Marion Boyars, 1985). Scientist Masaru Emoto studied how thoughts and intentions, as a form of language, could shape the physical world by altering the structure of water molecules. Masaru Emoto, *The Hidden Messages in Water*, trans. David Thayne (New York: Atria Books, 2004).
4 Donna Haraway, *Staying with the Trouble: Making Kin with the Chthulucene* (Durham: Duke University Press, 2016).
5 K.M. Newton, "Roland Barthes: 'The Death of the Author'," in *Twentieth-Century Literary Theory*, ed. K. M. Newton (London: Palgrave, 1997), https://doi.org/10.1007/978-1-349-25934-2_25.
6 Jacques Derrida, *The Copula Supplement*, in *Dialogues in Phenomenology*, ed. Don Ihde and Richard M. Zaner (The Hague: Martinus Nijhoff, 1975), 7–48.
7 Edward O. Wilson, *Half-Earth: Our Planet's Fight for Life* (New York: Liveright Publishing, 2016).
8 William Blake, *The Marriage of Heaven and Hell* (Boston: John W. Luce and Company, 1906), www.gutenberg.org/files/45315/45315-h/45315-h.htm.
9 bill bissett, *Nobody Owns the Earth* (Toronto: House of Anansi Press, 1971).

Chapter 8 – Storying

1 Thomas King, *The Truth About Stories: A Native Narrative* (Toronto: Anansi Press, 2002).
2 Ben Okri, *A Way of Being Free* (London: Head of Zeus, 2014).
3 Salman Rushdie, *Luka and the Fire of Life: A Novel* (New York: Random House, 2011).
4 Muriel Rukeyser, *The Speed of Darkness* (New York: Random House, 2006).
5 Jeffry R. Halverson, "Why Story is Not Narrative," *Centre for Strategic Communication, Arizona State University*, 8 December 2011, https://csc.asu.edu/2011/12/08/why-story-is-not-narrative/.
6 Ronald Lejano, Madeline Ingram, and Helen Ingram, *The Power of Narrative in Environmental Networks* (Cambridge, MA: MIT Press, 2013).
7 Derek Gladwin and Naoko Ellis, "Storying Systems Change: Learning and Living with Complexity," *Systemic Practice and Action Research* 38 (2025), https://doi.org/10.1007/s11213-025-09732-3.
8 Barry Lopez, "Interview," *Poets and Writers* 22, no. 2 (1994).
9 Nicholas P. Money, "Hyphal and Mycelial Consciousness: The Concept of the Fungal Mind," *Fungal Biology* 125, no. 4 (2021): 257–9.

10 George Lakoff and Mark Johnson, *Metaphors We Live By* (Chicago: University of Chicago Press, 1980). See also Vyvyan Evans, *The Crucible of Language: How Language and Mind Create Meaning* (Cambridge: Cambridge University Press, 2015).

11 Gregory Bateson and Mary Cathine Bateson, *Angels Fear: Towards an Epistemology of the Sacred* (New York: Hampton Press, 2004).

12 Tyson Yunkaporta, *Sand Talk: How Indigenous Thinking Can Save the World* (New York: Harper One, 2020).

13 George Lakoff and Mark Johnson, *Metaphors We Live By* (Chicago: University of Chicago Press, 1980).

14 George Monbiot, *Out of the Wreckage: A New Politics for an Age of Crisis* (London: Verso, 2017).

15 Geoffrey Supran, Stefan Rahmstorf, and Naomi Oreskes, "Assessing ExxonMobil's Global Warming Projections," *Science* 379, no. 6628 (2023), https://doi.org/10.1126/science.abk0063.

16 Paul J. Zak, "Why Inspiring Stories Make Us React: The Neuroscience of Narrative," *Cerebrum* 2 (2015), www.ncbi.nlm.nih.gov/pmc/articles/PMC4445577.

17 Marco Iacoboni, *Mirroring People: The Science of Empathy and How We Connect with Others* (New York: Picador, 2008).

18 Michael Nicoll Yahgulanaas, *Flight of the Hummingbird: A Parable for the Environment* (Vancouver: Greystone Books, 2008).

Chapter 9 – Witnessing

1 As a response to those who have continual worry about always needing to be doing something about the planet's future, see Derek Gladwin and Naoko Ellis, "Doing 'Nothing' as an Ecological Act," *The Arrow: A Journal of Wakeful Society, Culture & Politics* 12, no. 1 (2025).

2 Bruno Latour, "Why Has Social Critique Run Out of Steam? From Matters of Fact to Matters of Concern," *Critical Inquiry* 30 (2004): 225–48, www.bruno-latour.fr/sites/default/files/89-CRITICAL-INQUIRY-GB.pdf.

3 Fredric Jameson, *Postmodernism, or, The Cultural Logic of Late Capitalism* (Durham: Duke University Press, 1992).

4 Rachel Carson, *Silent Spring* (Boston: Houghton Mifflin, 2002).

5 Naomi Klein, *The Shock Doctrine: The Rise of Disaster Capitalism* (New York: Metropolitan Books/Henry Holt and Company, 2007). For more on this topic, see Antony Loewenstein, *Disaster Capitalism: Making a Killing Out of Catastrophe* (London: Verso, 2017).

6 "World is 'On Notice' as Major UN Report Shows One Million Species Face Extinction," *UN News: Global Perspective Human Stories*, 6 May 2019, https://news.un.org/en/story/2019/05/1037941.

7 Wallace Black Elk and William S. Lyon, "Never Leave Room for Doubt," in *Black Elk: The Sacred Ways of a Lakota* (New York: HarperCollins, 1991), 105–99. Don Hill, "Listening to Stones: Learning in Leroy Little

Bear's Laboratory – Dialogue in the World Outside," *Albertaviews: The Magazine for Engaged Citizens*, 1 September 2008, https://albertaviews.ca /listening-to-stones/.

Chapter 10 – Futuring

1 Jiddu Krishnamurti and David Bohm, *The Ending of Time: Where Philosophy and Physics Meet* (New York: Harper One, 2014).
2 Vanessa Machado de Oliveira, *Hospicing Modernity: Facing Humanity's Wrong and the Implications for Social Activism* (Berkeley: North Atlantic Books, 2021).
3 Hartmut Rosa, *The Uncontrollability of the World* (London: Polity, 2020).
4 See, for instance, Benjamin K. Sovacool, *Visions of Energy Futures: Imagining and Innovating Low-Carbon Transitions* (New York: Routledge, 2019).
5 Imre Szeman, *Futures of the Sun: Struggle over Renewable Life* (Minneapolis: University of Minnesota Press, 2024).
6 Ilya Prigogine, *The End of Certainty* (New York: The Free Press, 1997).

Index